Analysis and Design of Structural Bonded Joints

Analysis and Design of Structural Bonded Joints

By

Liyong Tong

BSc (Eng.), MSc (Eng), PhD, MIEAust, MAIAA
Senior Lecturer, Department of Aeronautical Engineering
The University of Sydney, NSW, 2006 Australia

Grant P. Steven

BSc, D.Phil, FIEAust, FRAe
Lawrence Hargrave Professor of Aeronautical Engineering
Department of Aeronautical Engineering
The University of Sydney, NSW, 2006 Australia

KLUWER ACADEMIC PUBLISHERS
Boston/Dordrecht/London

Distributors for North, Central and South America:
Kluwer Academic Publishers
101 Philip Drive
Assinippi Park
Norwell, Massachusetts 02061 USA
Telephone (781) 871-6600
Fax (781) 871-6528
E-Mail <kluwer@wkap.com>

Distributors for all other countries:
Kluwer Academic Publishers Group
Distribution Centre
Post Office Box 322
3300 AH Dordrecht, THE NETHERLANDS
Telephone 31 78 6392 392
Fax 31 78 6546 474
E-Mail <orderdept@wkap.nl>

Electronic Services <http://www.wkap.nl>

Library of Congress Cataloging-in-Publication Data

A C.I.P. Catalogue record for this book is available
from the Library of Congress.

Printed on acid-free paper.

Printed in the United States of America

To my parents, my wife Hua
and my children Richard and Victoria

LIYONG TONG

To all my family and friends
who have let me be an engineer for forty years

GRANT P. STEVEN

Contents

Preface

Most structures consist of an assembly of a number of individual components that must be connected to form an integral load transmission path. These connections are often referred to as joints and can be achieved in a variety of forms, e.g. by bolting, riveting, or other forms of mechanical fastening, or by welding or brazing for connecting metallic elements, or by adhesive bonding. No matter what forms of connections are used in the structure, these joints are potentially the weakest points in the structure and the locations where a weight penalty may apply. Thus structural joints must be designed adequately to meet the specific design requirements.

Adhesive bonding represents one of the most important enabling technologies for developing innovative design concepts and structural configurations as well as exploiting new materials. The evolution of adhesive bonding technology, and its current knowledge base, was made possibly by the explosive growth in the adhesive applications in a great variety of industries over the past few decades. While it is easy for everyone to identify examples of adhesive bonding in the world around us, analysis and design of structural bonded joints represent one of the most challenging jobs in structural design and manufacturing.

Compared to other joining methods, particularly mechanical fastening, adhesive bonding can offer substantial performance and economic advantages. For example, the ability to join dissimilar materials, to efficiently join thin-sheet materials. However, structural adhesives can not be used to directly substitute for rivets, bolts, welds or other fastenings. Joints originally designed using methods other than adhesive bonding must be thoroughly re-designed to make best use of the advantages of adhesive bonding while avoiding its limitations. To achieve this, an engineer must be fully aware of the advantages and the limitations of adhesive bonding, and must be have sufficient knowledge in mechanics of bonded joints and in polymer chemistry and material science related to adhesives.

This book aims to provide engineers and researchers with an up-to-date information relevant to analysis and design of structural bonded joints. In

Chapter 1, the reader is introduced to the concepts of bonded joints and adhesive testing standards. In Chapter 2, the fundamentals of stress analysis techniques are discussed from two aspects: the simplified analytical solution procedures and the detailed numerical solution methods. Chapter 3 is devoted to the prediction of joint strength, one of the most difficult tasks in joint design. This chapter discusses the strength prediction of joint from two approaches: one is based on strength of material criteria and the other is based on fracture mechanics. In Chapter 4, damage tolerance issues of bonded joints are addressed by discussing various possible damage and its effects on the stresses and strength of bonded joint. This provides a new angle to understanding failure mechanism of bonded composite joints. In Chapter 5, we review a number of methods of structural optimization, and then present some illustrative examples of optimized shapes of the adherend or spew fillet in bonded joints. Chapter 6 is devoted to a new branch of adhesive bonding technology, transverse stitching and stitched joints. Chapter 7 presents some selected applications of adhesive bonded joints in various industrial sectors, such as aerospace, marine, automobile etc.

ACKNOWLEDGEMENT

The authors would like to acknowledge the numerous researchers in the industrial and government organizations and at academic institutions who have contributed to the field of analysis and design of structural bonded joints and whose names do not appear in the bibliography included. The authors made every attempt to credit those individuals upon whose work the book is based. The authors are indebted to the Coorperative Research Center for Advanced Composite Structures (CRC-ACS), the Australian Research Council and the University of Sydney for supporting and funding much of the research from which this book was formed. We would like to acknowledge the benefits of frequent technical dialogues with close colleagues including: Associate Professor Don Kelly of the University of New South Wales; Drs Francis Rose, Alan Baker, Richard Chester, Chunhui Wang and Adrian Morris of the Aeronautical and Maritime Research Laboratory at the DSTO Australia; and Drs Lalit Jain and K.H. Leong of the CRC-ACS. The authors also would like to thank their former PhD student Dr Adrian Rispler for providing some of his research results.

Finally, we would like to thank our families for their support and acceptance of our dedication throughout this project, for without it, we could not have completed this task.

LIYONG TONG
GRANT P. STEVEN

1

Introduction

1.1 BACKGROUND

Adhesive bonding, either alone or in combination with other joining or reinforcing methods, represents one of the most important enabling technologies for developing innovative design concepts and structural configurations as well as exploiting new materials. An adhesive may be defined as a material which can be applied to surfaces of materials and can join them together to resist separation after consolidation. The materials being bonded together are normally referred to as adherends or substrates. Structural bonded joints may be considered as a class of joints that are formed by applying adhesives, either along or in combination with other joining or reinforcing methods, between two or more substrates or adherends, and have the ability to sustain relatively large externally-applied forces.

Most structures consist of an assembly of a number of individual components that must be connected to form an integral load transmission path. These connections are often referred to as joints and can be achieved in a variety of forms, e.g. by bolting, riveting, or other forms of mechanical fastening, or by welding or brazing for connecting metallic elements, or by adhesive bonding. No matter what forms of connections are used in the structure, these joints are potentially the weakest points in the structure and the locations where a weight penalty may apply. Thus structural joints must be designed adequately to meet the specific design requirements.

Compared to other more conventional methods of joining, particularly mechanical fastening, adhesive bonding can offer substantial performance and economic advantages. For example, the ability to join dissimilar materials, to join efficiently thin-sheet materials. However, structural adhesives can not be used to directly substitute for rivets, bolts, welds or other fastenings. Joints originally designed using methods other than adhesive bonding must be thoroughly re-designed to make best use of the advantages of adhesive bonding while avoiding its limitations. In doing so, an engineer must be fully aware of the advantages and the limitations of adhesive bonding, some of which are listed in Table 1.1 as a brief guideline.

Table 1.1. Advantages and limitations of bonded joints

Advantages	Limitations
Ability to form light weight, strong and stiff structures	Can not be disassembled
Ability to join dissimilar materials; e.g., metal to composite material	Residual stresses may be created due to difference in coefficients of thermal expansion
Ability to efficiently join thin-sheet materials	Limits to thickness joined with simple configurations
Improved stress distribution, no stress concentration in adherends, unlike mechanical fasteners which produce points of stress concentration	Sensitive to peel or through-thickness stress, i.e., a bonded joints generally is stronger under shear but weak in cleavage and peel
Enhanced fatigue properties due to improved stress distribution	Poor resistance to elevated temperature and fire
Improved corrosion resistance (sealed against corrosion); e.g. minimizing galvanic corrosion by interposing a non-conducting barrier between the metallic adherends	Prone to environmental degradation, uncertain long term durability under severe service conditions
Smoother surface finishes, with an absence of fastener heads, weld heads, etc.	Toxicity and flammability problems
Bonding process can be automated, most convenient and cost-effective joining; e.g. complex castings may be produced in several pieces and bonded together	Inspection can be difficult, quality control becomes critical, i.e. the effectiveness of adhesive bonding can not be fully assessed by inspection techniques currently available

Analysis and design of structural bonded joints is a multi-disciplinary task, and it can involve concepts from surface chemistry, polymer chemistry, stress analysis, and fracture mechanics as well as manufacturing technologies. When designing and building structural bonded joints, an engineer is faced with making decisions in the following important aspects:

(1) Selecting a suitable adhesive;
(2) Choosing appropriate surface preparation for the selected adherends;
(3) Developing an appropriate design based on rational stress and failure analysis;
(4) Deciding a quality-controlled fabrication process;
(5) Selecting an adequate protection of the joint from unacceptable hostile conditions in service;
(6) Consideration to inspection and repair.

This book is intended to focus only on the third item, i.e. various aspects of analysis and design of structural bonded joints, including stress analysis techniques, strength prediction, damage tolerance issues, optimum joint shape design, and bonded joints reinforced with transverse stitching. The rest of the five important aspects will be briefly discussed only and keen readers will be directed to the relevant references available.

1.2 JOINT CLASSIFICATIONS

Careful joint design is as important in bonded joints as other joining methods. It can often produce economic benefits by permitting the use of cheaper adhesives and simpler bonding process. An appropriate design of a structural bonded joint is clearly a quite onerous task, especially when the joint is designated to resist significant loads under various operational conditions. For instance, it is important to select or design an adequate and rational joint geometrical configuration. This selection is often based on a detailed information of stress distributions, properties of the selected materials as well as manufacturing cost consideration. It is not permissible to assume uniform stress distributions even for the simplest single lap joint configurations with relative short overlap. An adhesive often has low modulus and strength compared to metallic adherends. However, when it is used as a load transmission interlayer between two adherends, especially thin ones, its modulus and strength are more than adequate when selected appropriately.

Real structural bonded joints appear in various forms of geometrical configurations. Instead of discussing all possible forms of bonded joints and undertaking case studies, the text aims to provide up-to-date science and engineering information relevant to the fundamental nature of analysis and design of structural bonded joints. In doing so, some typical joint configurations will be discussed from the point of view of geometrical shapes, material selections and manufacturing processes.

1.2.1 Geometrical configurations

There are a large variety of bonded joint configurations in current engineering practice. Figures 1.1 and 1.2 show some typical classifications of joint being used in practice to carry a given set of loads. In Figure 1.1(a) to 1.1(i) are shown various lap joints that are commonly loaded in tension. Figure 1.1(j) to 1.1(l) shows the butt joint and two variations of butt joint design. Figures 1.2(a) to 1.2(c) depicts three types of Tee joints that are formed by connecting one vertical panel to a horizontal panel and are loaded in tension in the vertical panel. These types of Tee joints arc generally used in the ship building industry. Figures 1.2(d) to 1.2(g) show some typical joining arrangements between two or three tubes or pipes. Figures 1.2(d) to 1.2(f) show tubular joints between two tubes of different or identical diameters. Figure 1.2(g) shows a typical Tee type configuration of joining between one vertical and two horizontal tubes.

Any appropriate design of a structural bonded joint must be based on a reliable stress analysis and strength evaluation. It is a challenging task to adequately determine the stress distribution as it is complex even for the simplest single lap joint as shown in Figure 1.1(a). This is because the tensile loads in the single lap are not co-linear, and thus a bending moment is developed in the continuous adherend at either end of the overlap. The stresses in the adhesive are not only in shear but also in peel (the normal or tearing stress in the through-thickness direction). As adhesive is normally strong in shear and weak in peel, peel is the stress component that should always be avoided or minimized, if unavoidable. One of the effective methods of reducing the peel stress is to taper the end of an adherend as shown by the shaded triangular areas in Figure 1.1(a) to (f). When an adherend is tapered at its ends the local bending stiffness is then reduced and this results in a flexible adherend in terms of following the deformation of the continuous one.

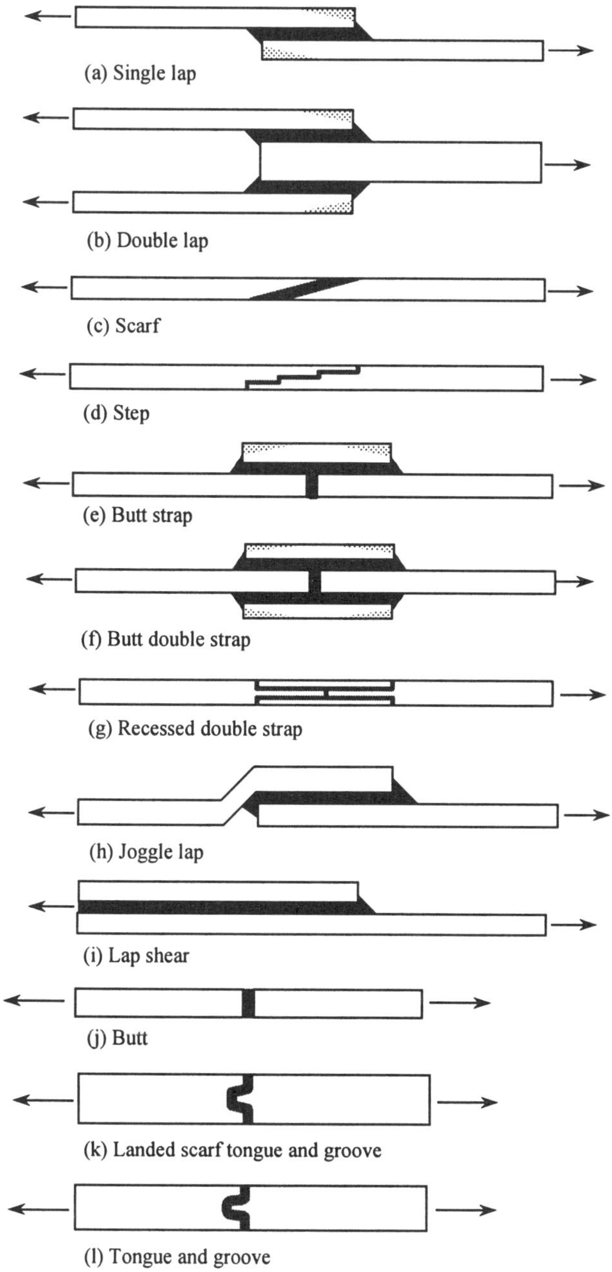

Figure 1.1 Some common structural bonded lap joints subjected to in-plane axial load, and a butt joint and two variations in butt joint design

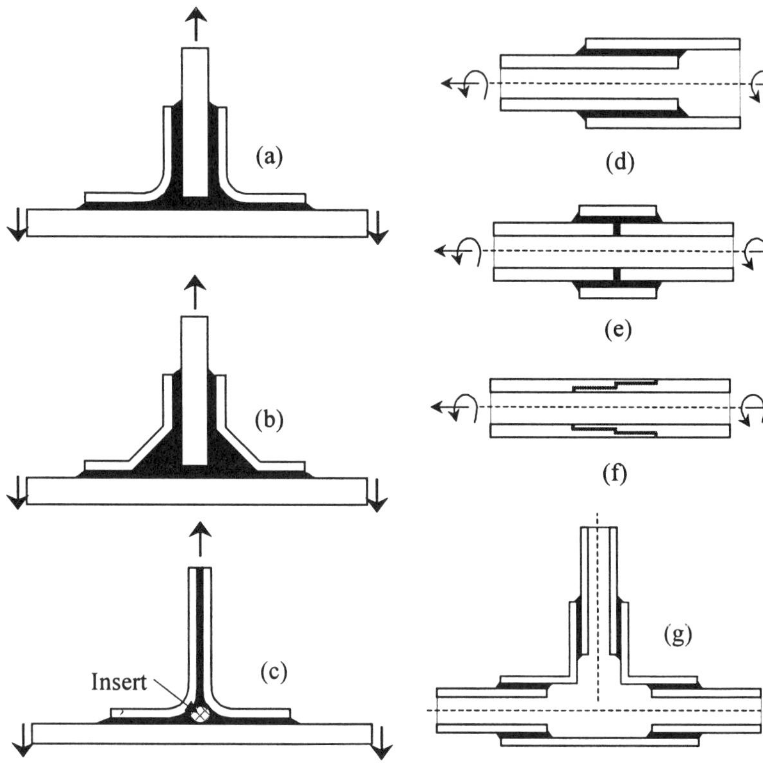

Figure 1.2 three typical Tee joints used in joining two flat panels, and some typical tubular joints used in bonding two or three tubes or pipes.

1.2.2 Material considerations

There are a variety of adhesives either organic or inorganic, e.g., animal glues, vegetable glues, mineral glues, elastomeric group of adhesive, thermoplastic and thermosetting adhesives. Thermoplastic adhesives may be repeatedly softened by heating and hardened on cooling without undergoing chemical changes, while thermosetting adhesives harden permanently on curing with molecular chains undergoing irreversible cross-linking. The reader is referred to Mays and Hutchinson (1992) and Adams, Comyn and Wake (1997) for further information.

Structural adhesives can be generally classified into the following groups: epoxy, polyester, Acrylic, polyurethane, polyimide. Amongst these adhesives, the most widely used structural adhesives are epoxies.

Epoxy adhesives typically contain several components, and the epoxide resin is the most important component. In practice, numerous epoxy adhesives can be formulated to achieve desirable properties by blending the base resins with a variety of materials, for example hardeners, flexibilisers, tougheners and fillers. Adhesives can come in a form of a thin film, and also can come in liquid form with several parts. Film adhesives can have a carrier. For example, FM300 film adhesives have three types of carriers, tight knit tricot, random mat and wide-open knit. In selecting an adhesive, an engineer should consider the following factors:

(a) Shelf life of the unmixed adhesive
(b) Viscosity, usable life, wetting ability and joint open time of the freshly mixed liquid adhesive
(c) Curing cycle and rate of strength development
(d) Mechanical properties, such as modulus, strength and fracture toughness, temperature and moisture resistance, creep and fatigue of hardened adhesive.

The materials that can be bonded together include timbers, metals, plastics, honeycombs and fiber reinforced plastics. The surface preparation processes may be different for different adhesives and adherend materials.

1.2.3 Manufacturing processes

Joints can also be classified from the point of view of manufacturing process into two categories: adhesive bonding and co-curing. Adhesive bonding is a process of joining two or more solid or hardened materials via application of adhesive layers. This is widely accepted and used in industry. Adhesive bonding has been used to join timbers, plastics, metals, as well as consolidated fiber reinforced composites. Adhesive bonding can be conducted at an elevated temperature or at an ambient temperature. Co-curing is a new concept widely accepted and used in joining composite structures. One important aspect of co-curing is that both adhesives and adherends are consolidated at the same time in one curing cycle. One example is to place film adhesive between composite plies and then to cure them in one curing process. Another example is to laminate a number of dry preform plies and then inject resin following either resin transfer molding or resin film diffusion process. With a stack of dry preform plies, transverse stitching in the through-thickness direction can also be applied to improve the interlaminar peel strength, particularly at that end of an overlap. Although, strictly speaking, it is not the same as the traditional adhesive bonding, stitching can form bonded joints in which adhesive layer can be regarded as reinforced with fibers. This is similar in concept to rubber toughened epoxies and to film adhesive with carrier. For this reason, chapter 6 has been included in this book to address this new concept of structural bonded joints or reinforced adhesive bonded joints.

1.3 ADHESIVE CHARACTERISATION

Any rational design of a structural bonded joint must be based on adequate knowledge of the stresses in the joint and the strength of the joint. To determine the stresses in a structural bonded joint and further to predict its strength and service life, it is necessary to know the material properties of adhesive and adherend. For a linear stress analysis, Young's modulus and Poisson's ratio are the two input data. For a material nonlinear analysis, stress-strain curves may be required and material yielding and hardening rules may also be needed. To predict static strength, material allowables or critical fracture toughness may be required. To determine service life, fatigue testing properties and effect of service conditions on adhesive properties must be obtained. In the following section, we will address the durability issue and environmental effects.

A great variety of standard test methods have been developed by government, industrial, society, and university investigators to characterize adhesive when sandwiched between different adherend materials. These test methods are continually updated and revised. Acceptable test methods are published in, for example, the ASTM standards (American Society for Testing and Materials), the BS standards (British Standards), the ISO standards (International Standards Organization), and the ES standards (European Standards). These tests not only evaluate the material properties of adhesive, such as strength, modulus and fracture toughness, but also evaluate bonding technique, effectiveness of surface preparation, and curing cycle etc. All tests can generally be classified into four groups, i.e., shear, tension, peel, and fracture toughness. Our intention is not to review all these test methods, not even to reference them all. The reader is referred to the books by A. J. Kinloch (1987) and by Adams, Comyn and Wake (1997) for quick reference and to the Standards available in most technical libraries for detailed information. In this book, a brief description of some commonly used tests from the four groups will be provided to illustrate some general techniques used for characterizing structural adhesives.

1.3.1 Shear tests

Shear tests are widely used to evaluate either the shear strength or the shear modulus and shear stress-strain curve of adhesives sandwiched between various adherends under various operating and environmental conditions. Shear tests are also useful for evaluating surface treatments and cure cycles by comparing apparent shear strength of joints made from a number of materials. Figure 1.3 schematically shows the geometrical configurations of common shear test specimens adopted in various standards.

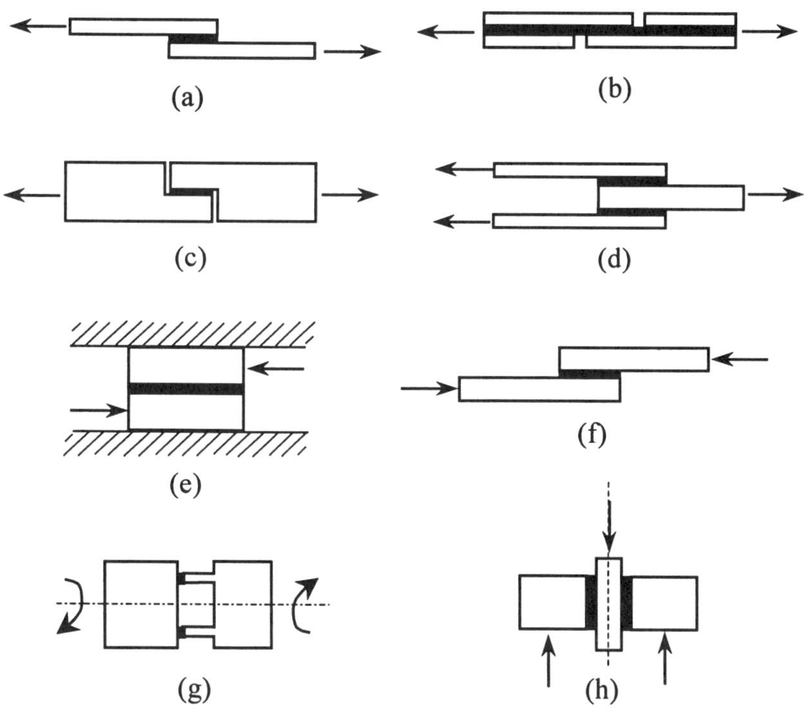

Figure 1.3 Specimen configurations of common shear tests: (a) single lap, (b) laminated assemblies, (c) thick adherend, (d) double lap, (e) modified rail in compression, (f) single lap in compression, (g) torsional shear (or napkin ring), and (h) pin-and-collar in compression.

Table 1.2 tabulates some of the ASTM and the BS shear test standards, which can be used to measure the apparent shear strengths, shear moduli and shear stress-strain curves of adhesives. The third column of this table lists the intended adherend materials, i.e., metal, plastic, rigid plastic, wood, composites, and fiber reinforced plastic (FRP).

Table 1.2 Shear test standards

Specimen	ASTM(or BS)	Substrate Materials
Single lap	D 1002-72(94)*	Metal
Fig. 1.3(a)	(BS 5350:C5 1976)	
	D 3163-73(96)	Rigid plastic
	D 3164-97	Plastic
	D 5868-95	Fiber reinforced plastics
Laminated assemblies Fig 1.3(b)	D 3165-73(95)	Metal, Plastic
Thick adherend	D 2339-94a	Wood
Fig 1.3(c)	D 3983-93	Wood, metal, Composites
	D 5656-95	Metal
Double lap	D 3528-76(96)	Metal
Fig 1.3(d)	(BS 5350:C5 1976)	
Modified rail	D 4027-81(93)	high density wood, metal, plastic
Fig 1.3(e)		
Single lap	D 905-49(94)	Wood
Fig 1.3(f)		
Torsional shear	E 229-70(97)	Metal
Fig 1.3(g)		
Pin-and-collar	D 4562-90(1955)e	Metal
Fig 1.3(h)		

*Number in parenthesis indicates the year in which the standard was re-approved.

All the shear tests listed in Table 1.2 are intended for determining the apparent shear strengths of adhesives between various adherend materials by applying tensile, compressive or torsional loading. Amongst these shear tests, single lap and double lap are two of the most commonly used specimen configurations because they are simple to construct and represent a close resemblance to the geometry and service conditions for many structural adhesives. However, the shear stress distribution in adhesive is not uniform although in almost all test methods it is conventional to define the apparent shear strength as the average shear stress in the bondline, e.g., the failure load divided by the bonding area. One should be aware of the fact that: (a) the maximum stress near the bondline ends may significantly differ from the average, and (b) the adhesive is not in pure shear stress state. Normal stress (peel) in the through-the-adhesive-thickness direction is almost inevitable, and it can change the failure mode and location depending on the geometry and materials of the adhesive and adherend. Comparing to the single lap shear test, double lap shear test has a low-peel stress distribution. It is common to represent this joint as essentially peel-free.

The shear modus of adhesives between various rigid adherends can be measured using the thick adherend shear test, the modified rail shear test and the torsional shear test. The ASTM D 3983-93 thick adherent shear test method is generally suitable for joints with rigid adherends and non-rigid adhesives. The ratio of the adherend tensile modulus to adhesive shear modulus should be greater than 300 to 1 and the adhesive shear modulus should be not greater than 700 MPa. The thick adherents can be wood, metal, or composites.

The shear stress-stain curve can be measured by using the thick adherend metal lap-shear test (ASTM D 5656) and the torsional shear test (ASTM E229). In the thick adherend shear test, the KGR-1 measuring device (Krieger, 1988) can be used to measure the relative axial movement which is then converted to the shear strain in adhesive. Thick adherend shear test specimen has a joint geometry simpler than the torsional shear specimen, and thus can be more easily made. It has been shown that the thick adherend geometry and materials could give non-uniform shear stress distributions for some adhesive thickness and stiffness (Kassapolou and Adelmann, 1991). This non-uniformity may introduce inaccuracies in the calculation of adhesive properties. The measured shear stress-strain curves of adhesive are imperative for carrying out accurate nonlinear stress analysis of adhesive bonded joints of various configurations. They also give the shear strength and the ultimate shear strain, which are critically useful for predicting the joint strength.

Adhesives respond differently in a small area joint versus a large area joint or in a coupon specimen versus a large structural component. This variation can be due to adhesive density, flow characteristics, cure rate, gel time, carrier composition, entrapped volatiles, volatiles released during cure, etc. and also due to cure cycle variables, such as temperature, time, pressure, rise rate to temperature, cool down rate etc. The laminated assemblies ASTM D 3165-73 (re-approved 95) describes a standard test method to measure the apparent shear strength of large bonded structural components.

For joining fiber-reinforced plastics (FRP) and metals, ASTM D 5868-95 describes a lap shear test for use in measuring the bonding characteristics of the adhesive. This test method is also applicable to random fiber oriented FRP. In addition, ASTM D5573-94 details the standard practice and method for classifying, identifying, and characterizing the failure modes in adhesively bonded fiber-reinforced-plastic (FRP) joints.

1.3.2 Tensile tests

Tensile tests are used for the determination of tensile modulus, tensile strength and tensile stress-strain curve of adhesives between substrates or adherends. The butt joint as shown in Figure 1.4.(a) is a typical specimen configuration. ASTM D 897-78 (95a) describes a standard test method for the determination of the comparative tensile properties of the adhesive bonds when tested on standard shape specimens and under defined conditions of pretreatment, temperature, and testing machine speed for wood and metal adherends. ASTM D 2095-72 (96e1) describes the standard test method for determination of the relative tensile strength of adhesives by the use of bar- and rod-shaped butt-joined specimens under defined conditions of preparation, conditioning, and testing. This test method is applicable to the testing of adhesives with various adherend materials in either similar or dissimilar combinations. ASTM D2094-69 (91) describes standard procedure for preparing bar and rod specimens for adhesion tests. It has been shown that the tensile stress distribution can be non-uniform due to mismatch of Poisson's ratio and possible misalignment of the two bars (Adams and Coppendale, 1979). Figure 1.4 (b) shows a sandwich type butt joint described in ASTM C297-61. Figure 1.4(c) shows a cross-lap type of joint specimen described in ASTM D1344-78. This cross-lap assembly can be used to determine the comparative strength of adhesives.

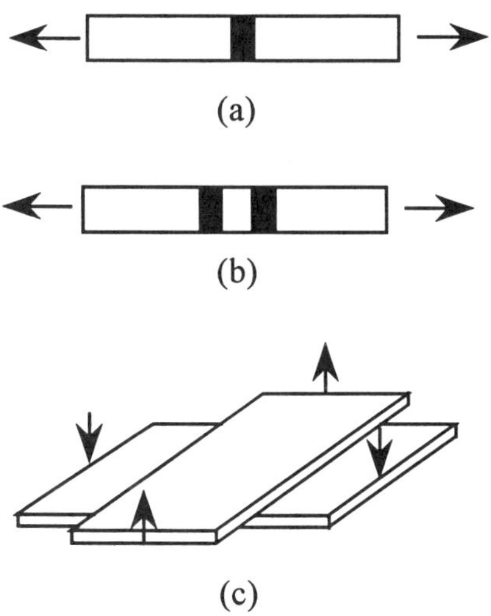

Figure 1.4 Specimen configurations of common tensile tests: (a) butt joint, (b) sandwich butt joint, and (c) cross-lap joint

1.3.3 Peel tests

Peel tests are extensively used to determine the peel resistance or force of adhesive bonds between (a) relatively flexible adherends; and (b) a relatively rigid adherend and a relatively flexible adherend. Figure 1.5 depicts schematically the specimen configurations of common peel tests. In a peel test, a thin flexible strip that is bonded to a substrate is pulled apart at some angle to the underlying substrate. Figure 1.5(a) depicts a 180° peel test described in the ASTM D 903-49 (93) and the British Standards (BS) BS 5350: C11 (1979). This test method can be used to determine the comparative peel or stripping characteristics of adhesive bonds when tested on standard-sized specimens and under defined conditions of pretreatment, temperature and testing machine speed. Figure 1.5(b) shows schematically the climbing drum peel test described in the ASTM D 1781-76 (93). This test can be used to determine the peel resistance of adhesive bonds between: a relatively flexible adherend and a rigid adherend; and the relatively flexible facing of a sandwich structure and its core, when tested under specified conditions. Figure 1.5(c) shows the T-peel test defined by the ASTM D 1876-72 (95) and the BS 5350:C12 (1979). This test method is primarily intended for determining the relative peel resistance of adhesive bonds between flexible adherends by means of a T-type specimen. Figure 1.5(d) shows the 90° peel test described in the British Standards BS 5350: C14 (1979) and BS 5350: C10 (1976). Figure 1.5(e) shows the floating roller peel test configuration as described in the ASTM D 3167-76 and the BS 5350: C9.

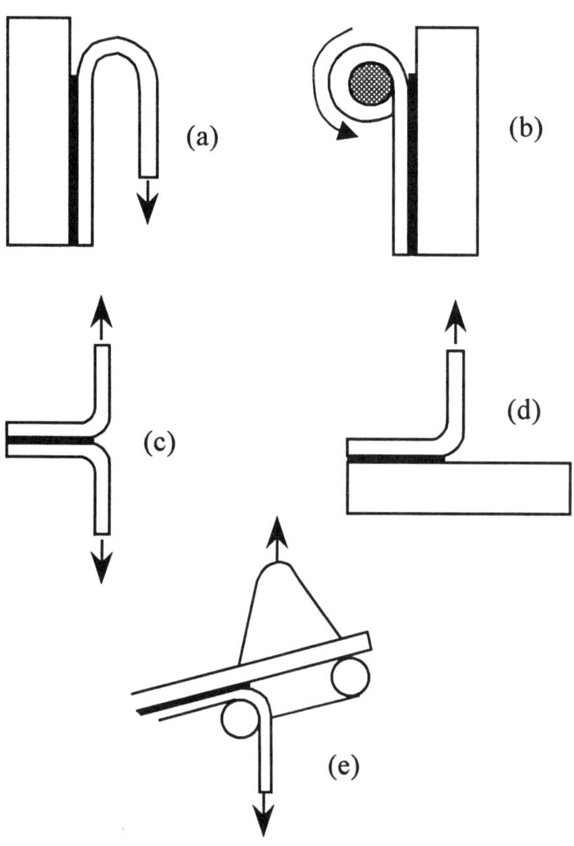

Figure 1.5 Specimen configurations of common peel tests: (a) 180° peel, (b) climbing drum peel, (c) T peel, and (d) 90° peel, (e) floating roller peel.

Recently the peel test has found applications in microelectronics industries and is being used to study the adhesion of thin metal films on dielectric substrates. There exists a research effort devoted to clarify and define the mechanisms of adhesion between these two dissimilar materials.

1.3.4 Fracture toughness tests

Adhesive joints often fail by a crack propagating either in the bondline or along the interface between a substrate and the bondline or in a mixture form of both. Thus fracture mechanics has been widely used to correlate crack growth behavior in adhesive joints. Fracture toughness tests are used to determine the fracture toughness or energy release rates of adhesives sandwiched between substrates. These properties of adhesive bonds are important in terms of characterizing adhesives, predicting adhesive joint strength and service life, and designing rationally adhesive joints. Figure 1.6 shows the three crack propagation modes, they are opening mode (mode I), shearing mode (mode II) and tearing mode (mode III). A crack in the bondline of adhesive joint is constrained and will propagate under mixed-mode conditions in a direction such that the crack tip is in pure mode I (Fleck et al, 1991).

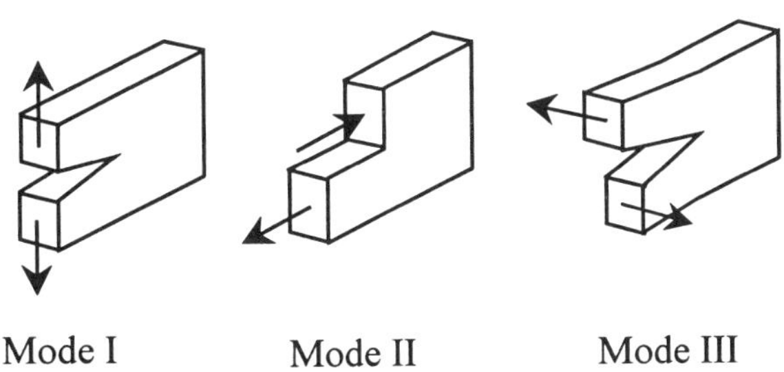

Figure 1.6 Three crack-propagation modes: mode I opening mode, mode II shearing mode, and mode III tearing mode.

Figure 1.7 depicts some common standard test configurations for mode I. Figure 1.7(a) –(c) are specimen configurations that can be used to measure the cleavage strength or the mode I fracture toughness. ASTM D 1062-78 (96e1) describes the standard test method for determining the comparative cleavage strength of adhesive bonds sandwiched between metallic materials (see Figure 1.7(a)). ASTM D 3807 –79 (93) describes a standard test method for determining the cleavage/peel strength of adhesive for bonding engineering plastics (see Figure 1.7 (b)). The tapered DCB test configuration shown in Figure 1.7 (c) is described in the ASTM D 3433-75 (93). This test method can be used to determine the fracture strength in cleavage of adhesives bond, which is useful in terms of developing design parameters for bonded assemblies. This test method is intended for use in metal-to-metal applications although it may be used for measuring fracture properties of adhesives sandwiched between plastic adherends.

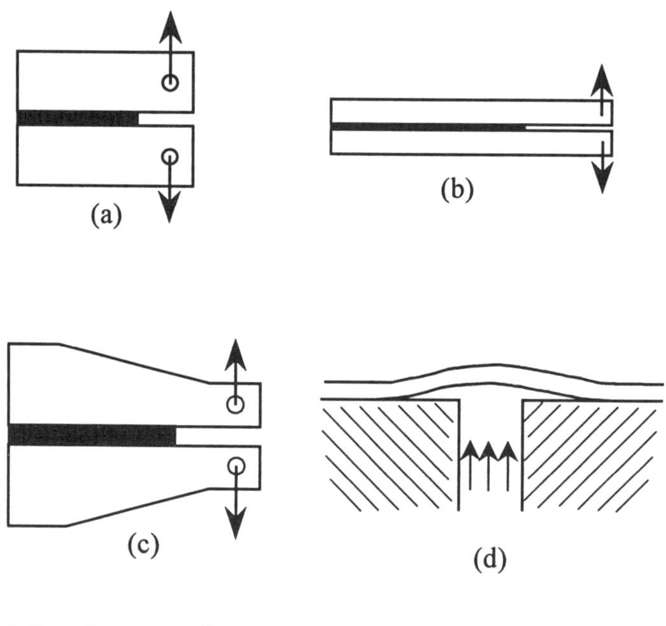

Figure 1.7 Specimen configurations of common fracture toughness tests: (a) compact tension, (b) double cantilever beam (DCB), (c) Tapered DCB, and (d) blister test.

Figure 1.7 (d) shows the blister test described in the ASTM D714-87 (1994)e1. This test method employs photographic reference standards to evaluate the degree of blistering that may develop when paint systems are subjected to conditions that will cause blistering. While primarily intended for use on metal and other nonporous surfaces, this test method may be used to evaluate blisters on porous surfaces, such as wood, if the size of blisters falls within the scope of these reference standards. When the reference standards are used as a specification of performance, the permissible degree of blistering of the paint system shall be agreed upon by the purchaser and the seller.

There exist other types of fracture toughness test methods for measuring Mode II fracture toughness, e.g. end-notched flexure (ENF) test, and for measuring in-situ mixed mode fracture toughness of mode I and mode II. A test to accomplish this has been offered by Reeder and Crews (1991) known as the mixed-mode bending (MMB) test. Figure 1.8 shows schematically the ENF and the MMB test specimens. The ENF specimen has been used to measured mode II fracture toughness of adhesive bonded joints by many researchers (Chai, 1993a). The relative locations of loading and supporting for the MMB specimen as well as the joint geometry can be varied using a special loading jig to measure the entire range of mixed mode fracture toughness (Ferlund and Spelt, 1994b).

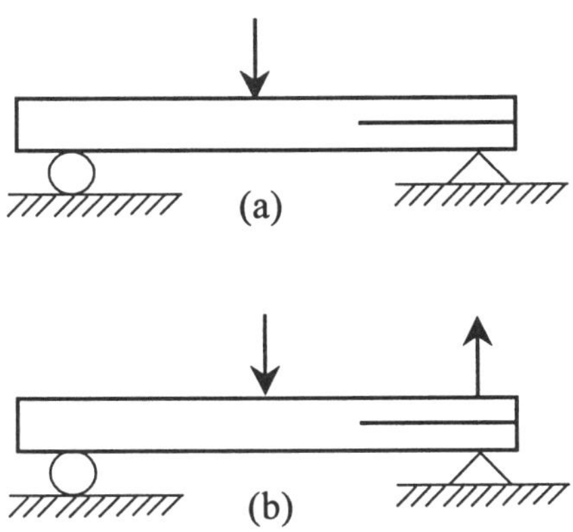

Figure 1.8 Schematic configurations of ENF and MMB specimens: (a) ENF specimen, (b) MMB specimen.

1.4 FATIGUE AND DURABILITY

The estimation of service life of an adhesive joint represents one of the most challenging and complicated aspects in analysis and design of structural adhesive joints. In service, the joint may be subjected to a dynamic or cyclic loading, and may be exposed to various environmental conditions, such as temperature, moisture, relative humidity, chemical agents, oxygen etc. The service life expected could be in a very large time-scale, and the service conditions could vary constantly. These factors make it impossible to evaluate the service life of a joint via accurately simulating the operating conditions or impractical via even approximately applying the service conditions. However, the service life of an adhesive joint may be estimated in carefully controlled or exposed laboratory environments under a specific controlled loading. These laboratory tests usually represent a fraction of the service life with an increased applied loading comparing to that in service, and are therefore regarded as accelerated tests. Although the laboratory tests may not be easily done with confidence, they are widely used to estimate the fatigue life of adhesive joints under various environmental conditions, and to compare the fatigue performance of various adhesives.

Table 1.3 tabulates some standard test methods of adhesive bonded joints under various service conditions, such as temperature, moisture, creep, impact, cyclic loading, water etc. Figure 1.9 shows the geometrical configurations of the block shear and wedge test specimens as listed in Table 1.3.

Characterization of fatigue properties is a time-consuming, expensive and challenging task, and plays a very important role in predicting fatigue strength of bonded joints. One of the most important fatigue parameters to

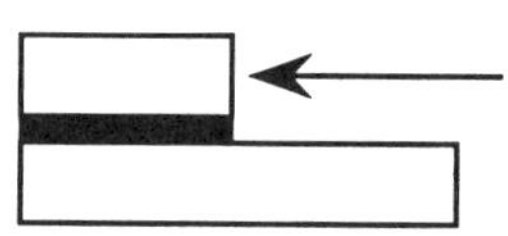

(a)

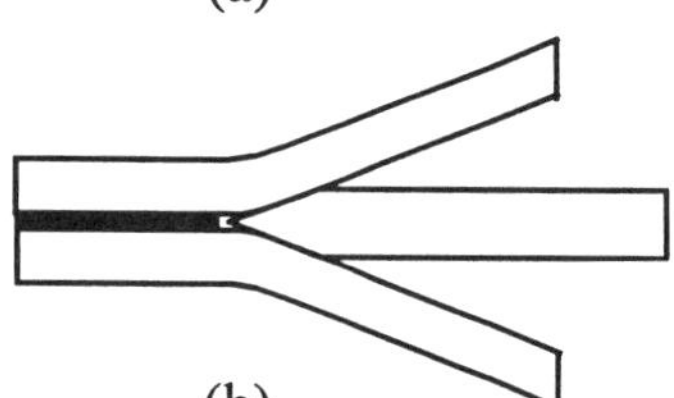

(b)

Figure 1.9 Specimen configurations: (a) Impact resistance, (b) Wedge test.

Table 1.3 Test standards of adhesive joints under various service conditions

Specimen	ASTM (or BS)	Service conditions
Single lap Fig 1.3a	D2295-72 (96)*	Shear strength at high temperature 315 to 850°C (600 to 1500°F)
	D2557-72 (93)	Shear strength at subzero temperature –267.8 to -55°C (-450 to –67°F)
	D1780-72 (94)	Creep under temperature, tensile shear stress and time
	D2294-69 (96)	Creep under tensile stress and temperature –55 to 260°C (-67 to 500°F)
	D2293-69 (96)	Creep under compressive stress and temperature,
	D4680-92	Creep under stress and temperature and humidity
	D3166-73 (93)	Fatigue in metal bonds
Block shear Fig 1.9a	D950-82 (94)	Impact resistance
Lap shear	D2919-95	Durability under stress and environmental and temperature exposure
T peel	D2918-93	Durability under stress and environmental and temperature exposure
Wedge test Fig 1.9b	D3762-79 (93)e1	Surface durability
	D1151-84(90)	Exposure to moisture and temperature
	D1183-96e1	Cyclic laboratory aging conditions
	D3632-77 (90)e1	Oxygen degradation in wood bonding
	D1879-70 (87)	High energy radiation
	D904-94	UV light aging
	D896-97	Chemical reagents
	D1828-91	Atmospheric exposure

*Number in parenthesis indicates the year in which the standard was re-approved.

be determined is the number of cycles of an alternating loading to produce failure and its sensitivity to loading frequency and amplitude as well as combined service conditions, such as temperature and moisture. Another important parameter is the steady state fatigue growth rate and its dependent relationship to loading frequency and amplitude. Standard test methods, such as ASTM E 647-88 (95a), E 1761-95 etc, may be used to process the testing data and determine the steady state fatigue growth rate. When the fatigue growth rate is very low, the fatigue cracks or damages simply stop growing, and the joint may have an infinite service life.

The creep properties one should consider in adhesive bonding include the time-dependent deformation or strain and the time to failure when subjected to static or quasi-static loading. These properties can be characterized using standard test methods, for example, those given in Table 1.3, and will serve as an another important design factor. For example, some structures may not allow large creep deformation between the two bonded components.

As pointed out in Table 1.1, one of the major limitations of adhesive bonding is the poor resistance to elevated temperature and the susceptibility to environmental degradation. When the service temperature is sufficiently raised, the strength and Young's modulus are significantly reduced while the ultimate strain and deformation are dramatically increased. This is because a structural adhesive performs best when used at a temperature below its glass transition temperature. When adhesive bonded joints operate in an environment of elevated temperature and are subject to moist or immersed in water, the adhesive bonding performance can be significantly degraded. The most confidence would be gained by exposure of the bonded structures to their real operating environments for the whole of their expected service lifetime. However, this is impractical. Test joints would have to be exposed to an aggressive or accelerated operating environment in laboratories, such as high humidity and elevated temperature. There are many factors affecting the durability of a bonded joint, such as type of alloy used as adherends, surface treatment methods, salted water and mechanical stresses. Thus the operating temperature range and the service environments play a key role in deciding whether to use adhesive bonding or other joining techniques by considering adhesives and surface treatment methods available.

2

Stress analysis techniques

2.1 INTRODUCTION

Stress analysis is one key step followed by engineers in any rational structural design. It should provide an engineer with reasonably accurate information on distribution of stress or strain in the real structures subjected to specified loading and service conditions. This information will then enable the engineer to predict the strength and service life of the designed structures. This is also true in any bonded joint design. Accurate determination of stress distribution in adhesive bonded joints represents one of the most challenging structural stress analysis problems. This is because bimaterial interfaces and geometric discontinuities create stress concentrations and uncertainty in material behaviors arises. To make joint design more difficult, any appropriate joint should not be a fuse or weak link in any structure. There exists a large amount of information available on stress analysis and strength prediction of adhesive bonded joints. This chapter and the following chapter are intended to provide an up-to-date overview of the methods on stress analysis and strength prediction.

This chapter will focus on discussion of some stress analysis techniques for some typical joint configurations and experimental validations of the predicted stress distribution. Chapter 3 will describe various failure criteria based on strength of materials and fracture mechanics and their applications to joint strength predictions. It is certainly not all inclusive, but should provide an insight to up-to-date methods on analysis and design of adhesive bonded joints.

Stress analysis techniques can generally be classified into two major categories, i.e., analytical analysis based heavily on a number of mathematical assumptions or simplifications, and numerical analysis using finite element method (Zienkiewicz and Taylor, 1989) on a digital computer. Analytical analysis is normally conducted based on a number of simplified assumption in order to obtain fast and simple solutions with reasonable accuracy, while numerical analysis is conducted using a commercially available general purpose software and can provide further detailed stress distribution. Both methods complement each other. Analytical results can rapidly provide an engineer with some aspects of the stress distribution that may be used in quick joint design by conducting parametric study, for example, selection of adhesives, fundamental geometry etc. While numerical analysis can provide detailed stress information, it requires more effort and it is best used to fine tune up the joint configuration. Numerical analysis can also be combined with an optimization procedure to achieve an optimal joint design, which will be further discussed in Chapter 5.

It's not an easier job to calculate the stresses in an adhesive bonded joint than to predict budding stress of thin walled structures of contact problems. To start a stress analysis for a bonded joint, one needs to define the joint geometry, to have sufficient knowledge of the material properties and to know the loading and the operating conditions. While the shape of an adherend can be easily defined according to a design, description of the accurate shape of an adhesive layer or bondline presents some difficulty. For example, a geometrical description of the adhesive layer or bondline is uncertain or becomes very complex. The geometry of the adhesive layer includes bondline thickness and bondline end shapes or spew fillet. In typical bonding process, adhesive in a liquid state is consolidated into a solid state. While the bondline thickness may be controlled using metal shims or spacers or carrier mats, the shape control of the spew fillet produced by overflow of liquid adhesive at a bondline end becomes quite difficult if not impossible. An accurate geometry of the adhesive spew fillet cannot be measured if no measure was taken in preparing the specimen. Another imperative issue is determination of adhesive mechanical properties. Although there are many standard test methods for measuring adhesive properties, adhesive manufacturers are not currently capable of supplying the appropriate mechanical properties that are directly useful to stress analysis engineers.

2.2 ADHESIVE PROPERTIES

Adhesives may be characterized using the up-dated standard test methods following two ways: one is the characterization of bulk adhesive, and the other is determination of in-situ adhesive properties in adhesive-bonded joints. The specimens and test methods for plastic materials can be used to characterize the bulk adhesives.

The bulk adhesive properties are intrinsic and not influenced by the adherends. In the case of adhesive joints adhesives are sandwiched between adherents and their properties are subjected to the influence of adherends and are thus more representative of real behavior. An immediate question that can be asked is; are the bulk adhesive properties the same as those of adhesives sandwiched between adherents?

For a while, it was believed that the bulk adhesive properties are different from the adhesive properties in adhesive bonded joints due to an improper interpretation of carefully designed tests (Adams and Coppendale, 1977). Jeandrau (1986, 1991) has confirmed that there exists a good correlation between the adhesive properties in bulk form and the properties of adhesive sandwiched between adherends. Table 2.1 lists five types of adhesives tested and Table 2.2 tabulates the comparison of the measured shear modulus G using the thick adherent test method and the bulk adhesive shear modulus G predicted using the measured Young's modulus E and the following equation

$$G = \frac{E}{2(1+\nu)} \tag{2.1}$$

where the Poisson's ratio ν varies between 0.3 and 0.5.

Table 2.1 Types of structural adhesive investigated (after Jeandrau, 1991)

Adhesive	Manufacturer	Adhesive type and curing process
A-Redux 322	Ciba-Geigy	Modified epoxy film (supported), curing at 175°C, 1h
B-DP 70	Evode	One-part, modified epoxy paste, curing at 190°C during 30 min
C-AV 119	Ciba-Geigy	One-part, modified epoxy paste, curing at 150°C during 1 h
D-A 2840	CECA SA	One-part, modified epoxy paste, curing at 200°C during 1 h
D-9323 B/A	3M	Two-part, modified epoxy paste, curing at room temperature (24 h) + post-curing (1 h at 80°C)

Table 2.2 Predicted bulk shear modulus G_B/experimental in-situ shear modulus G_C (after Jeandrau, 1991)

Adhesive	Measured tensile Young's modulus (bulk adhesive) E_T (MPa)[a]	Predicted shear modulus using Equation (2.1) (bulk adhesive) G_B (MPa)		Measured shear modulus (bonded specimens) G_C(MPa)[b]
		ν=0.5	ν=0.3	
A	4500 ± 100	1500	1730	1660 ± 220
B	3302 ± 63	1100	1270	1029 ± 68
C	3386 ± 133	1129	1302	1058 ± 53
D	5000 ± 115	1667	1923	1600 ± 75
E	2156 ± 95	719	829	650 ± 100

[a] Tensile test.
[b] Thick adherend test using the extensiosmeter developed by Althof and Brockmann (1976)

Comparison of the shear moduli of various adhesives in Table 2.2 reveals that there is a good agreement between the shear modulus measured in the bulk adhesive form and that measured in the in-situ sandwiched forms. It should be borne in mind that the quality of bulk adhesive specimens without defects, such as voids and porosity, is vital as the contents and size of defects in the thin adhesive layer sandwiched between adherends are generally smaller than those in the bulk adhesive specimens.

Table 2.3 compiles the mechanical properties or mechanical property ranges of some typical adhesives and structural materials from various sources. The adhesive material data are for reference only, and readers should measure the required adhesive properties following the standard test methods as briefly described in Chapter 1 for the purpose of structural bonded joint design.

Adhesives can be classified into two categories: one is a brittle adhesive and the other is a ductile adhesive. For a brittle adhesive, a proportional linear relationship exists between the stress and strain, while for a ductile adhesive, a nonlinear stress-strain relationship is generally observed (see Figure 2.1a). For brittle adhesives, Young's modulus and Poisson's ratio are the two adhesive properties that are required to perform a linear stress analysis for bonded joints. However, for ductile adhesives, the entire nonlinear adhesive behavior must be adequately determined and then taken into account in the stress analysis in order to accurately predict stresses and strains in an adhesive bonded joint, and further to predict its service life.

Table 2.3 Comparison of the mechanical properties of adhesives (Adams, Comyn and Wake, 1997) and structural materials (Gere and Timoshenko, 1991). The adhesive data are for illustrative purpose only.

Materials	Density (kg/m^3)	Young's modulus (GPa)	Poisson's ratio	Tensile strength (MPa)	Coefficient of thermal expansion (10^{-6}/°C)
Aluminum alloy	2600-2800	70-79	0.33	100-550	23
Brass	8400-8600	96-110	0.34	200-620	19.1-21.2
Glass	2400-2800	48-83	0.20-0.27	30-1000	5-11
Steel	7850	190-210	0.27-0.30	400-1200	10-18
Titanium alloy	4500	100-120	0.33	900-970	8-10
Adhesives					
Acrylic	1190	1.1-1.4		20-50	
Epoxy	1200	1-4		50-80	
Polyimide	1420	2-3		70-150	

The nonlinear adhesive behavior may be modeled with a measured true stress-strain curve, either in pure tension or in pure shear, and a mathematical model that takes all stress combinations into account. Figure 2.1b depicts a typical true shear and tensile stress-strain curves of FM 300k film adhesive measured using thick adherend test and butt joint test (Tong, 1997).

Gali, Dolev and Ishai (1981) found that the bulk adhesive properties obtained by a uni-axial test, such as tension, compression and torsion, can be related to the properties of an 'in-situ' adhesive layer in shear by a combined stress law that follows a modified Von Mises failure criterion. This criterion can give a function that relates the tensorial stress or strain components to their respective values. It is shown that this function is valid throughout the entire range of stress/strain relationship up to failure and under all combinations of compound stresses.

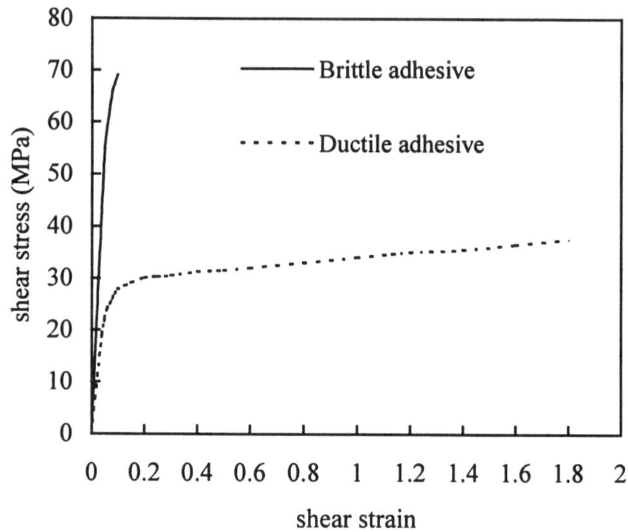

Figure 2.1a Comparison of shear stress-strain characteristics for brittle and ductile adhesives

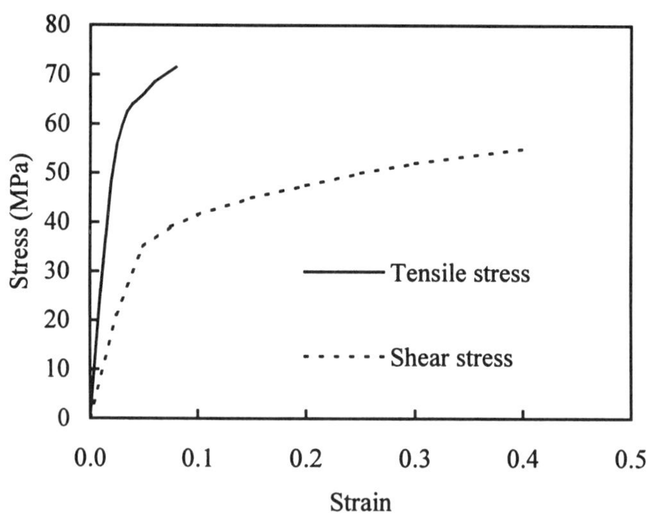

Figure 2.1b True shear and tensile stress-strain curves of FM-300 film adhesive

Jeandrau (1991) neglects the volumetric component of the stresses and gives

$$\sigma_c = \sigma_T = \sqrt{3}\tau$$
$$\varepsilon_c = \varepsilon_T = \sqrt{3}/2(1+2)\gamma \qquad (2.2)$$

where σ_c, σ_T are the tensile and compressive stresses, ε_c and ε_T are the tensile and compressive strains. τ and γ are the shear stress and strain. These equations indicate that once we measure the shear stress-strain curve, the tensile stress-strain relationship can be established. Therefore, only the shear stress-strain curve will be discussed here. Another reason for this is that shear stress-strain curve can be used directly in classical simple solution based on shear lag model.

The measured stress-strain curves can be characterized by a variety of mathematical models for the sake of analytical and numerical analysis. Figure 2.2 depicts some of the existing mathematical models. Hart-Smith (1973a) chose the following two special types of shear stress-strain behaviors for adhesives:

(a) Ideal elastic-plastic model for shear stress-strain behavior:

$$\tau = \begin{cases} G\gamma & \gamma \le \gamma_e \\ \tau_p & \gamma_e \le \gamma \le \gamma_p \end{cases} \qquad (2.3)$$

where γ_e, γ_p and G are chosen in such a way that the maximum strain energy density in shear, computed using the ideal elastic-plastic model, is equal to that computed using the measured shear stress-strain curve (see Figure 2.2a).

(b) Bilinear model for shear stress-strain behavior:

$$\tau = \begin{cases} G_e\gamma & \gamma \le \gamma_e \\ \tau_e + G_p(\gamma - \gamma_e) & \gamma_e \le \gamma \le \gamma_p \end{cases} \qquad (2.4)$$

where γ_e, γ_p, G_e and G_p are chosen in the same way as that used for the elastic-plastic model (see Figure 2.2a).

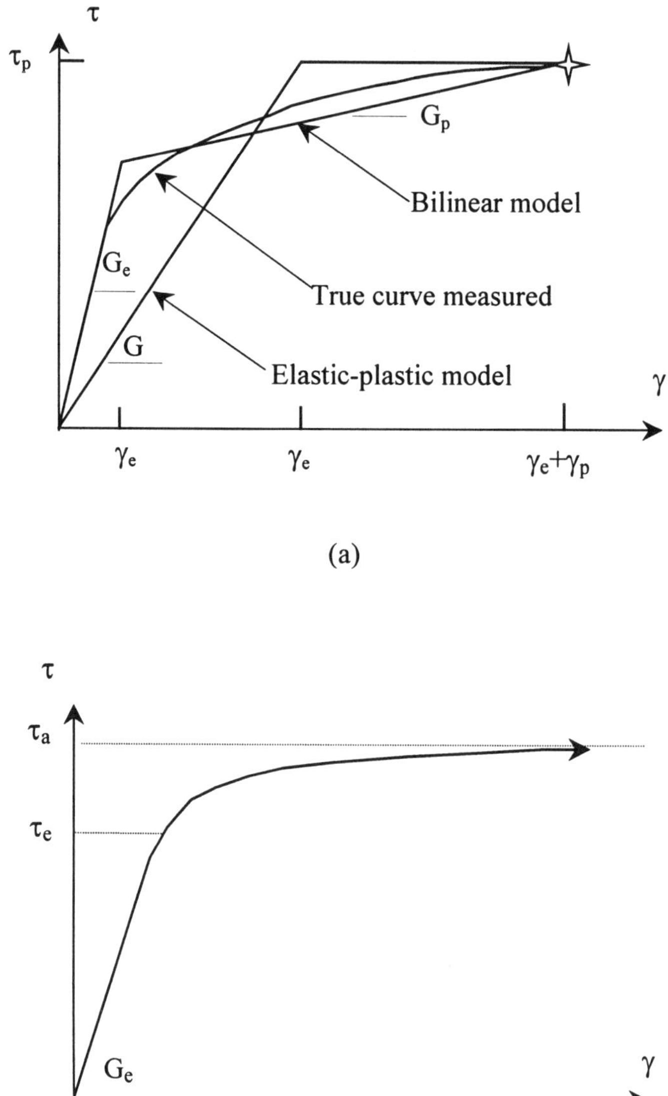

(a)

(b)

Figure 2.2 Models of shear stress-strain curves for adhesive. (a) Idealized elastic-plastic models used by Hart-Smith (1973a), (b) Grant model (ESDU, 1979)

The bilinear model may be closer to the true adhesive characteristic over its entire range of strain, thus a single model may be used for calculating the stresses. In the elastic-plastic model, if the maximum stress is less than yield, the true elastic curve may be used, while if the peak stress is between yield and failure stresses, a different and more accurate model is chosen. Hence the bilinear model is different from the elastic-plastic model as it is not necessary to first calculate the stresses to establish which intermediate elastic-plastic model should be used.

Another adhesive model used in analyzing the inelastic shear stress and strain in the adhesive bonded lap joints loaded in tension or shear (ESDU, 1979) is shown in Figure 2.2b, and the following expressions are utilized to represent the shear stress-strain characteristic

$$\tau = \begin{cases} G_a\gamma & \gamma \le \gamma_e \\ \tau_e + \dfrac{\alpha\beta}{\alpha+\beta} & \gamma_e \le \gamma \end{cases} \tag{2.5}$$

where

$$\alpha = G_a\gamma - \tau_e, \qquad \beta = \tau_a - \tau_e$$

τ_e and γ_e are the proportional elastic shear stress and strain respectively, while τ_a is the maximum shear stress.

While these mathematical models can be described by simple formulas, which in turn make it possible to develop analytical solutions, it is argued that selection of the model or models for a given adhesive represent a compromise between precision and convenience.

Several factors, which affect the measurement of adhesive properties e.g. stress-strain curves or modulus, include the test methods, the operational conditions such as temperature, moisture, loading rate etc. As there is a large amount of information in the literature on this issue, only a brief summary will be given here.

The shear properties of adhesives are measured primarily using the following three test methods, the napkin ring test, the Althof thick adherend test (Althof and Brockmann; 1976) and the KGR-1 thick adherend test (Krieger, 1988). Althof et al (1979) measured shear stress-strain curves of adhesive FM-73 at room temperature using the above three test methods. The curves are schematically plotted in Figure 2.3 to demonstrate the effect of testing methods on the shear stress-strain curve of adhesive FM-73 (Data taken from Althof et al, 1979).

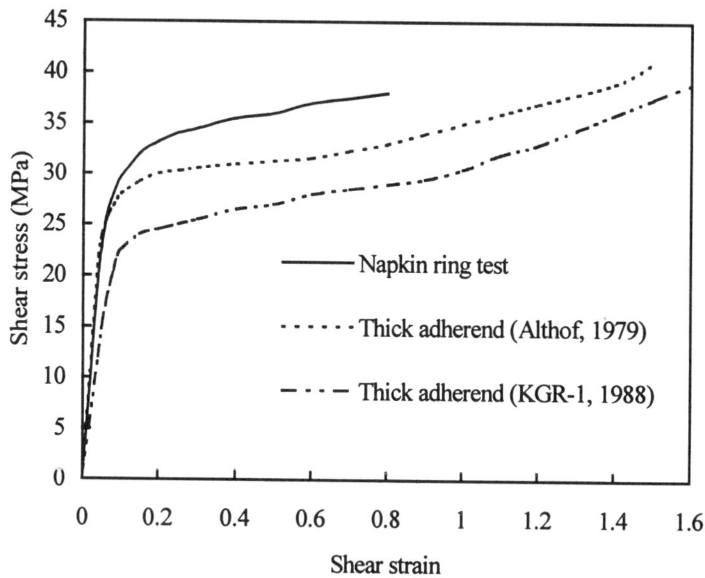

Figure 2.3 Schematics of shear stress-strain curves measured using different test methods (Data taken from ESDU, 1981)

Operating environmental conditions, such as temperature and moisture, have a critical effect on the behavior of adhesives and their stress-strain curves. Figure 2.4 depicts the measured shear stress-strain curves of a ductile adhesive subjected to a room temperature, -55°C and +70°C. Clearly, the adhesive becomes less ductile when tested in a cold environment, and more ductile when subjected to a hot environment. The elastic limit decreases as the environmental temperature increases. To make matters worse, relative humidity or moisture can also have a detrimental effect on adhesive behavior. The two extremes, i.e., cold and dry environment, and hot and wet environment, should be considered when assessing adhesive properties and designing adhesive bonded joints. The Engineering Sciences Data Unit has presented some interesting test results of various adhesives (ESDU, 1981). In addition, the rate of loading can have an important effect on shear stress-strain curves of adhesives. A fast loading rate tends to produce an increase in the elastic limit while a slow loading rate causes a decrease in the elastic limit. A pause during a load control test is likely to produce a creep step, and a stress relaxation is likely to be noticed during a pause in a displacement control test. It is desirable to conduct test at a constant displacement or strain rate rather than a constant load rate.

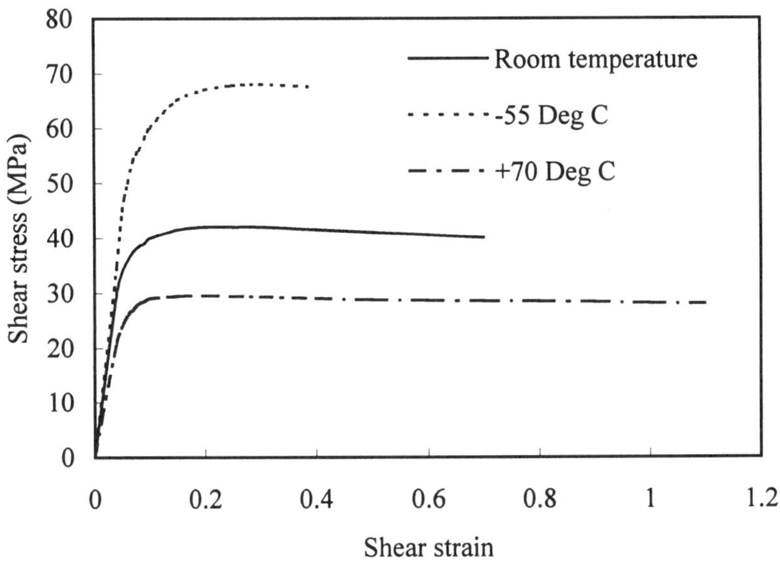

Figure 2.4 Schematics of typical shear stress-strain curves of the adhesive measured using thick adherend test at various temperatures (Data taken from ESDU, 1981)

2.3 ANALYTICAL ANALYSIS

Lap joint is one of the most common joints in practice, and the single lap joint has been widely used in a variety of standard tests for evaluating adhesive performance and quality control. Single lap joint test is simple and economical to conduct, and it also represents a close duplication to the loading type experienced by structural adhesives in service. For this reason, single lap joint has been used in a variety of standard tests as outlined in Chapter 1. For example, ASTM D 1002 describes a standard test method using single lap joint as shown in Figure 2.5. In this test, an approximate cross head speed of 1.3 mm (0.05 in) per minute is recommended. The apparent stress at failure is defined as the failure load divided by the bonding area, i.e.,

$$\tau = \frac{P}{bl}$$

This represents the simplest stress analysis of a single lap joint, and assumptions of this simple analysis are the adherends are rigid, i.e., do not exhibit elastic deformation; and the adhesive deforms only in shear. This apparent shear strength may be used to compare performance of various adhesive sandwiched between metal adherents under various service condition. The constant average shear stress does not represent the real stress distribution in the bondline due to the oversimplified assumptions.

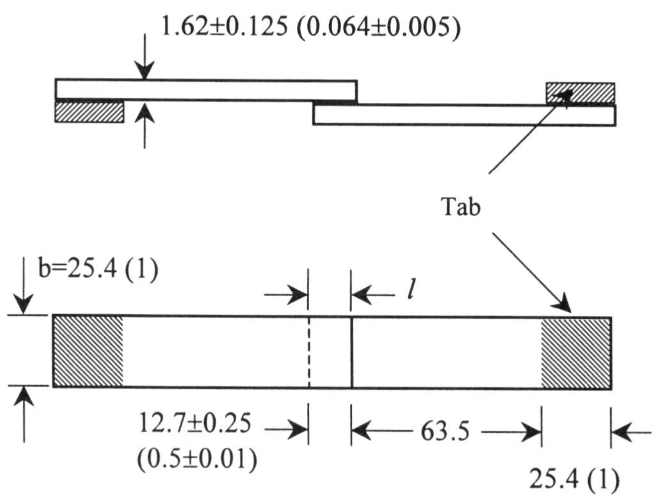

Figure 2.5 ASTM D 1002 standard single lap joint configuration. All dimensions are in mm (for convenience dimensions in inch are given in parenthesis)

2.3.1 One dimensional analysis for single lap joint

The stress analysis of the single lap joint has been under continuous development and refinement for the past six decades. In the following, a brief summary of the current results will be presented.

(a) Volkersen analysis (1938)

Volkersen (1938) presented a theoretical analysis of lap joints by considering only the stress arising from the differential staining in the lap joint as schematically shown in Figure 2.6. In other words, the rigid adherend assumption was abandoned, and the adherends are allowed to deform elastically. The differential straining between the adherends result in a non-uniform shear stress distribution in the bondline with a maximum at each end of the overlap. Defining the stress concentration factor η to be the ratio of the maximum shear stress to the mean stress, Volkersen found that

$$\eta = \frac{\delta}{\varepsilon}\left\{\frac{2\varepsilon^2 - 1 + \cosh(2\varepsilon\delta)}{\sinh(2\varepsilon\delta)}\right\} \tag{2.6}$$

where

$$\delta^2 = \frac{2c^2 G_a}{E_2 t_2 \eta} \qquad \varepsilon^2 = \frac{E_1 t_1 + E_2 t_2}{2E_1 t_1} \tag{2.7}$$

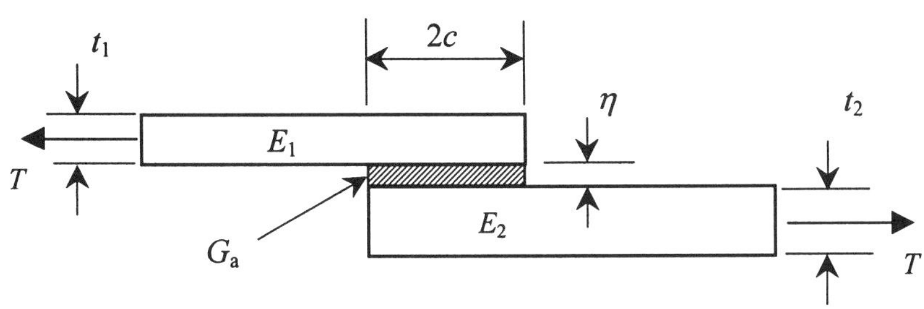

Figure 2.6 The single lap joint considered by Volkersen (1938)

When both adherends have the same thickness t and Young's modulus E, $\varepsilon^2=1$ and the stress concentration factor becomes

$$\eta = \delta \coth \delta \tag{2.8}$$

where

$$\delta^2 = \frac{2c^2 G_a}{Et\eta} \tag{2.9}$$

(b) Goland and Reissner analysis (1944)
The Volkersen's analysis based on a shear lag model is unsatisfactory in the sense that it does not include the peel stress in the bondline due to the loading eccentricity. Goland and Reissner (1944) formulated a theory that includes the peel stress. The undeformed single lap joint considered by Goland and Reissner is depicted in Figure 2.7(a). In Figure 2.7(a), α_n represents the force-eccentricity angle, $2c$ the overlap length, l the free length of the adherend, t the adherend thickness, and η the bondline thickness. The elastic modulus and the Poisson's ratio are E and ν for the adherends, and E_C and ν_C for the adhesive. When the joint is subjected to a tensile load T, there exist the direct force N, edge shear force V_0 and the edge bending moment M_0 all per unit width acting in the continuous adherend at the overlap end as depicted in the free-body diagrams in Figure 2.7(b). Since the eccentricity angle, $\alpha_n = t/2(l+c)$, is usually much smaller than unity (i.e., <<1), the direct force N can be approximated as

$$N = T\cos\alpha_n \cong T = pt \tag{2.10}$$

where p is the average applied tensile stress.

The self-equilibrium condition in the overlap portion as shown in Figure 2.7(b) can be used to express the shear force V_0 in terms of the edge bending moment M_0 as

$$V_0 = \frac{T(t+\eta) - 2M_0}{2c} \tag{2.11}$$

It is clear that the edge moment M_0 can not be determined by considering equilibrium conditions only.

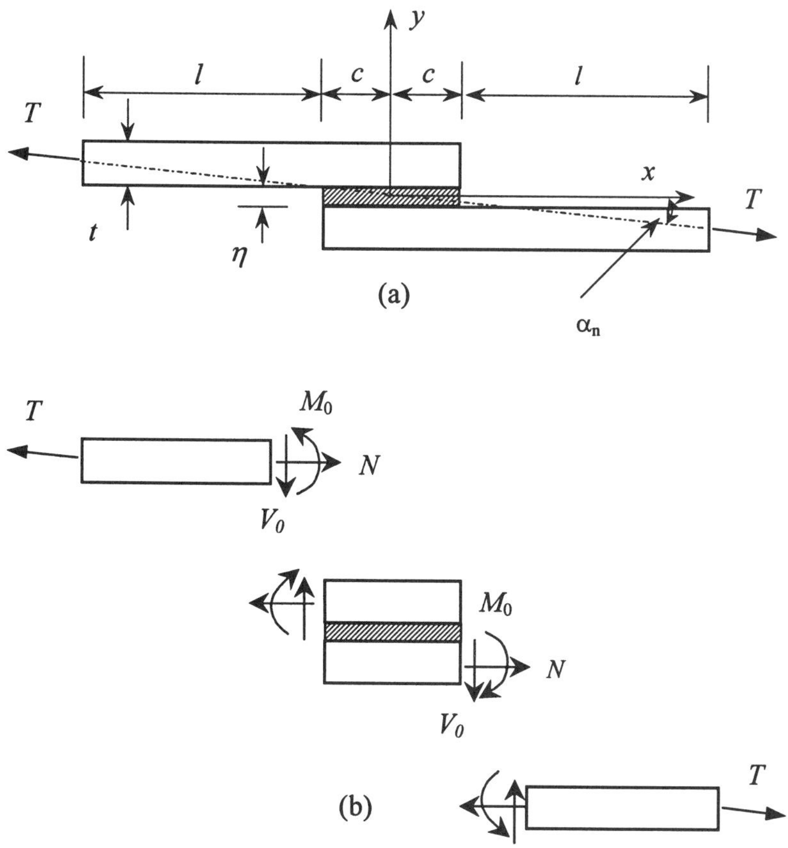

Figure 2.7 (a) an undeformed single lap joint considered by Goland and Reissner (1944), and (b) free-body diagrams subjected to applied load T

Goland and Reissner (1944) assumed that the adhesive layer be neglected because it is very thin compared to the adherend thickness ($\eta << t$). Thus E_C, ν_C and η were eliminated from their formulation. Large deflection of the free adherends was taken into account in their formulation. The simply supported boundary conditions were imposed together with the continuity conditions of the deflection and the rotation at overlap ends. The bending stiffness of the overlap was calculated by lumping together the upper and lower adherends. The edge moment M_0 was determined as

$$M_0 = \kappa \frac{Tt}{2} = \frac{\kappa p t^2}{2} \tag{2.12}$$

and

$$\begin{aligned} \kappa &= \frac{1}{1 + 2\sqrt{2}\coth(u_1 l)\tanh(u_2 c)} \\ &\cong \frac{1}{1 + 2\sqrt{2}\tanh(u_2 c)} \end{aligned} \tag{2.13}$$

where

$$u_1 = \sqrt{\frac{12\left(1 - \nu^2\right)T}{Et^3}}, \qquad u_2 = \frac{u_1}{2\sqrt{2}} \tag{2.14}$$

Note the edge moment M_0 (or the moment factor k) is dependent of E, ν, t, c, l and T or approximately depends only on E, ν, t, c and T. The approximation in equation (2.13) holds when l/c ranges from 5 to 100. When l/c is less than 1, $coth(u_1 l)$ can not be approximated as one (Tsai and Mortan, 1994).

For joints with relatively flexible adhesive ($E_C/E \le \eta/10t$), it was assumed that: (a) both adherends in the joint overlap undergo cylindrically bending only, and (b) the shear and peel (transverse normal) stresses in the relatively thin adhesive layer are constant. By using the values of M_0, V_0 and N given in the above equations, the adhesive shear and peel (transverse normal) stresses (τ_0 and σ_0) were obtained as

$$\frac{\tau_0}{p}\frac{c}{t} = -\frac{1}{8}\left\{\frac{\beta c}{t}(1 + 3k)\frac{\cosh\frac{\beta c}{t}.\frac{x}{c}}{\sinh\frac{\beta c}{t}} + 3(1 - k)\right\} \tag{2.15}$$

$$\frac{\sigma_0}{p}\left(\frac{c}{t}\right)^2 = \frac{1}{\Delta}\left[\begin{aligned} &\left(R_2\lambda^2\frac{k}{2} + \lambda k'\cosh\lambda\cos\lambda\right)\cosh\lambda\frac{x}{c}\cos\lambda\frac{x}{c} \\ &+\left(R_1\lambda^2\frac{k}{2} + \lambda k'\sinh\lambda\sin\lambda\right)\sinh\lambda\frac{x}{c}\sin\lambda\frac{x}{c} \end{aligned}\right] \tag{2.16}$$

where

$$\beta^2 = 8\frac{G_C}{\mathrm{E}}.\frac{t}{\eta}$$

$$\gamma^4 = 6\frac{\mathrm{E}_C}{\mathrm{E}}.\frac{t}{\eta'}, \qquad \lambda = \gamma\frac{c}{t}$$

$$R_1 = \cosh\lambda\sin\lambda + \sinh\lambda\cos\lambda$$

$$R_2 = \sinh\lambda\cos\lambda - \cosh\lambda\sin\lambda$$

$$\Delta = \frac{1}{2}(\sinh 2\lambda + \sin 2\lambda) \tag{2.17}$$

and

$$\frac{2M_0}{pt^2} = k, \qquad \frac{V_0 c}{pt^2} = k' \tag{2.18}$$

It is worth noting that the expression for the peel stress σ_0 in equation (2.16) is the original one developed by Goland and Reissner. Sneddon (1961) pointed out the inconsistency of the sign of the edge shear force V_0 in the Goland and Reissner's formulation of the boundary conditions, and then obtained an expression similar to equation (2.16) but with negative signs for the κ' terms involving the edge shear force V_0. Subsequently, many researchers have used Sneddon's expression. However, Carpenter (1989) found that the original expression for the peel stress by Goland and Reissner is correct. Tsai and Morton (1994) also pointed out that the original expression for the peel stress developed by Goland and Reissner is actually correct as the effect of the first inconsistency of the sign of the edge shear force V_0 was removed by a second sign inconsistency. They found that the original adhesive stress solutions of Goland and Reissner (1944) are accurate enough to predict the adhesive stress distributions in the global sense for the short and long joints. For bonded single lap joint, Wang and Rose (1997) provides an analytical solution for the triaxial stresses, i.e., the transverse shear and peel stresses and the two normal stresses in the length and width directions.

As an illustrative example, consider an single lap joint, as shown in Figure 2.7, which has the following parameters: E=70 GPa, ν=0.3, E_c=2.4 GPa, ν_c=0.33, η=0.15 mm, t=1.5 mm and c=30 mm. The applied axial load T is assumed to be 400 N/mm. Figure 2.8 depicts the curves of the nondimensional shear and peel stresses in the adhesive versus the nondimensional coordinate measured from the middle point of the overlap for the single lap joint. Clearly both the shear and peel stresses attain their maximum values at the overlap end.

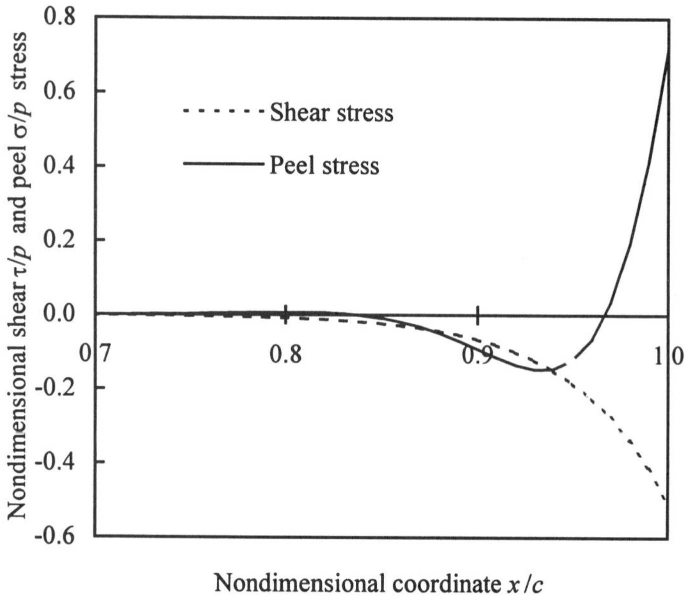

Figure 2.8 Shear and peel stress distribution in adhesive layer using equation (2.15) and (2.16)

(c) Hart-Smith analysis (1973b)

One of the assumptions made by Goland and Reissner was that the bondline is so thin that the presence of the adhesive doesn't affect the determination of the edge bending moment M_0. Hart-Smith coupled the determination of the edge bending moment M_0, and the adhesive shear and peel stresses, τ_0 and σ_0, by considering the individual deformations of the upper and lower adherends. The edge bending moment M_0 obtained by Hart-Smith is a function of all the material and geometry parameters. However, after neglecting small terms and applying the approximation of $\sinh u_1 l \approx \cosh u_1 l \approx e^{u_1 l}/2$, the edge bending moment M_0 can be simplified as

$$M_0 = k\frac{Tt}{2} \cong T\left(\frac{t+\eta}{2}\right)\frac{1}{1+u_1c+\frac{u_1^2c^2}{6}} \tag{2.19}$$

The Hart-Smith analysis determines the edge moment M_0 and the adhesive stresses simultaneously, and takes into account the effect of large deflection of the free adherends, but ignores the large deflection effect in the joint overlap.

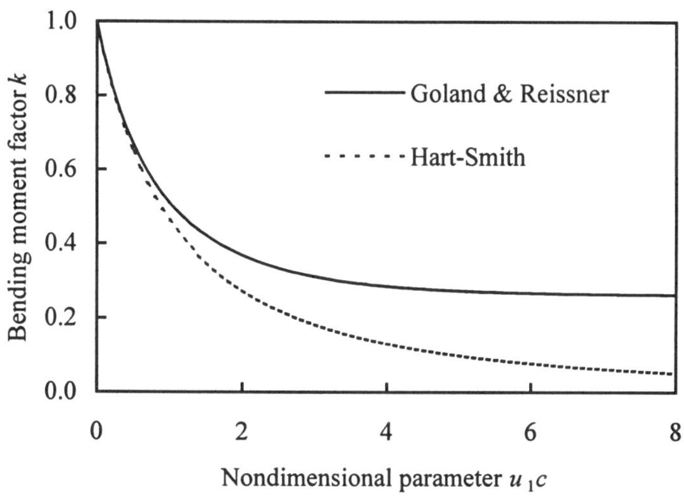

Figure 2.9 Bending moment factor κ versus u_1c

Figure 2.9 plots the edge bending moment factor k versus the nondimensional parameter u_1c for the two solutions given by Goland and Reissner (equation 2.13) and Hart-Smith (2.19). The nondimensional parameter u_1c is proportional to the overlap length when the material properties and the applied load are given, and represents the level of the applied load when the material properties and joint dimensions are fixed. When parameter u_1c is small, say for example less than one, both solutions give almost the same edge bending moment factor. However, when u_1c is larger than one, the difference between the two solutions becomes significant.

(d) Oplinger analysis (1991)
A further development was conducted by Oplinger (1991) who took into account the large deflection effect in the joint overlap and presented a more detailed analysis. According to Oplinger's formulation (1991), the edge moment M_0 is given by

$$M_0 = \frac{Tt}{2} \frac{R_3(1+R^2C_2)+8R_4\frac{T_{h21}}{T_{h22}}R(C_1-C_2)}{R_3+8R_4\frac{T_{h21}}{T_{h22}}RC_1+\sqrt{8}(1+R^2C_1)\frac{T_{h21}}{T_{h1}}} \tag{2.20}$$

where

$$R=\sqrt{24\frac{p\eta(1+\nu_C)}{tE_C}}, \qquad R_3=\sqrt{\frac{8\nu_1}{R}}, \qquad R_4=\sqrt{\frac{\nu_2}{8}},$$

$$\nu_1,\nu_2 = a \pm b, \qquad a=[4(1+\frac{3\eta}{4t})+\frac{R^2}{4}], \qquad b=\sqrt{a^2-R^2},$$

$$T_{h2j}=\tanh(\mu_j\frac{c}{t}) \quad (j\text{=1,2}), \qquad T_{h1}=\tanh(u_1 l),$$

$$\mu_1=\frac{R_3}{\sqrt{8}}R\sqrt{\frac{E\eta}{G_C(1-\nu^2)t}}, \qquad \mu_2=\sqrt{8}R_4\sqrt{\frac{E\eta}{G_C(1-\nu^2)t}},$$

$$C_1=\frac{K_1}{K_2}, \qquad C_2=\frac{1}{K_2},$$

$$K_1=\frac{12}{16R_4^2-R^2}, \qquad K_2=\frac{48R_4^2}{4R_4^2-1} \tag{2.21}$$

The edge bending moment factor given by Oplinger (1991) is between the Goland and Reissner's solution and the Hart-Smith's solution. Tsai and Mortan (1994) conducted a 2-D nonlinear finite element analysis and discussed the limits of the above three solutions for the edge bending moment in the range of short, intermediate long to long overlap.

2.3.2 One dimensional analysis for other joint configurations

Analytical solutions for other joint configurations have been developed under various assumptions by many researchers in the past few decades. For example, solutions for double lap joints, butt joint, lap shear joint, stepped or scarf joint and tubular joints. Due to space limitation, only a few examples will be discussed in this section.

Figure 2.10(a) depicts schematically a lap shear joint subject to in-plane end loading. A lap shear joint may be representative of many applications in aerospace structures, such as stiffened panels and shells, selective plate reinforcement, bonded edge doublers for flush mechanical attachments etc. The lap shear joint has only one end with critical stress concentration, and it has long been known that the doubler must be flexible in bending at the end to diffuse load transfer gradually. This is different from a single lap joint, which has two overlap ends with stress concentrations, although both joint configurations share many of the governing equations. Hart-Smith (1973d) has presented an elastic-plastic analysis for lap shear joints, which is referred to as doubler in his study, with a range of end support conditions.

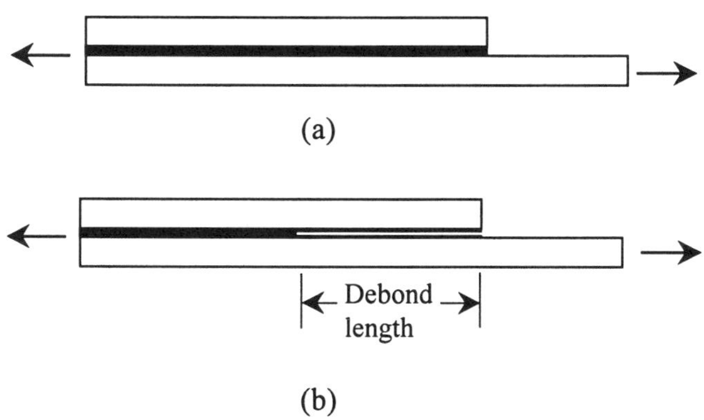

(a) (b)

Figure 2.10 Lap shear specimen of equal thickness

To investigate fracture of bonded joints under mixed mode at one overlap end only, the lap shear joint is particularly useful. Fracture mechanics behavior can be investigated by placing an initial crack in the bondline and forming a cracked lap shear joint (see Figure 2.10(b)), which was originally proposed by Brussat et al (1977). The presence of a crack has then been used to study adhesive joint debonding by many researchers. An ASTM round robin was conducted and reported by Johnson (1987) on stress analysis of cracked lap shear joints. Edde and Verreman (1992) proposed a stress analysis using an approach similar to that suggested by Goland and Reissner (1944). In their analysis, both adherends are assumed to have the same thickness and elastic modulus. These assumptions decouple the normal and shear stress solutions. In a recent study, Lai et al (1996) revisit the cracked lap shear joint specimens and introduce a new nondimensional closed form solution for the lateral deflection and bending moment. They considered three types of boundary conditions, clamped-clamped, roller-clamped and roller-roller, as it was found that the testing results are sensitive to boundary conditions (Lin and Liechti, 1987). The clamped-clamped case is representative of typical testing in fixed grips of a universal testing machine, while the roller-roller case is equivalent to half of the single lap joint problem solved by Goland and Reissner (1944).

Another joint configuration that received considerable attention is the double lap joint, as shown in Figure 1.1b. Hart-Smith (1973a) studied adhesive bonded double lap joints using elastic-plastic analytical techniques. Explicit solutions obtained include sufficiently simple formulas for predicting the shear bond strength and the plastic zone length. It was shown that for a given double lap joint with a particular geometric configuration and specified material properties including the ideal elastic-plastic or the bilinear adhesives the maximum bond shear strength achievable between specified adherends can be characterized by the adhesive shear strain energy per unit bonded area. Hart-Smith's formula is limited to the elastic-plastic and bilinear adhesive models. Similarly, ESDU (1979) pointed out that the precise shape of the shear stress-strain curve has no effect on the limiting joint strength, and can affect only the adhesive shear stress distribution along the overlap. Once again, a particular adhesive model is utilized in ESDU's analysis. Tong (1994) proposed a study of adhesive bonded double lap joints with an arbitrary nonlinear adhesive shear stress-strain behavior.

Step and scarf joints (see Figure 1.1 c and d) have also been investigated by many researchers because they are representative of actual structures. Step or scarf joints have also been used in composite repair for a smooth surface finish. Lubkin (1957) presented a theory for calculating the elastic stresses in scarf joints. Erdogan and Ratwani (1971) determined the shear stress distribution in the adhesive layer and the axial stresses in the two adherends for stepped or scarf joints. In their analysis, both adherends are assumed to undergo longitudinal deformation only and adhesive layer is modeled by pure shear springs, thus their analysis is based on a shear lag model. Hart-Smith (1973c) included adhesive plasticity and derived continuum mechanics solutions for the static load-carrying capacity of scarf and stepped lap joints.

The 90° peel test (see Figure 1.5d) has been extensively used to measure adhesion strength. In the test, a thin flexible strip that is bonded to a substrate by an adhesive layer is pulled apart at an angle of 90° to the underlying substrate. The peel force required to separate the strip from the substrate is related to the adhesion strength and has been widely used for joint design purposes. Spies (1953) presented the first elastic theoretical analysis of the mechanics of the 90° peel test by modeling the bonded part as an elastic beam on an elastic (Winkler) foundation and the flexible part as being elastic. In the absence of plastic deformation, the peel force is a direct measure of the adhesive fracture energy. However, when plastic deformation takes place, e.g. in the flexible strip, the plastic dissipation and the residual energy left in the adherend must be considered in order to determine the fracture energy from a measured peel force. There exist several elastic-plastic analyses of the peel test (See, Chang et al, 1972; Gent and Hamed, 1977; Kim and Aravas, 1988). In these analyses, elastic-plastic deformation of the thin strip is considered.

One dimensional analyses of adhesively bonded joint of two tubes of different diameters (see Figure 1.2d) have also been carried out by various researchers because joining of these tubes becomes critical in the design stage for an efficient load-carrying structure. Lubkin and Reissner (1956) investigated the distribution of stress in the adhesive lap joints between two thin circular cylindrical tubes subject to axially distributed loads. For these tubular joints subjected to torsional loads, Volkersen (1965) gave a closed form solution for the shear stress in the adhesive layer. Alwar and Nagaraja (1976) considered the adhesive as a viscoelastic material and calculated the stress distribution in the adhesive layer for an adhesively bonded tubular joint in tension. Shi and Cheng (1993) presented an analytical analysis of adhesive-bonded cylindrical lap joints subject to an axial load. Adams and Peppiatt (1977) analyzed the stresses in adhesive bonded tubular lap joints subjected to axial and torsion loads using axi-symmetric finite element models. In the axial loading case, the stress concentrations arose by three mechanisms: differential straining, bending introduced by the non-co-linearity of the overlapping tubes and the end effects. Ikegami et al (1995) examined the method of tapering the adhesive coupling at the boundary of the cylinder joint (fillet) as compared to a straight edge cylinder coupling at the boundary of the lap joint. Matsui (1990) investigated the effects of geometric size and mechanical properties of both adherends and adhesive on the average ultimate shear stress of the bonded rectangular and tubular lap joints. Chon (1982) analysed composite tubular lap joint in torsion and developed a closed form solution for the stress distribution. More comprehensive viscoelastic analyses of adhesive bonded tubular joints under torsion were given Zhou and Rao (1993).

2.3.3 Effect of nonlinear adhesive behavior

(a) Hart-Smith and ESDU

Ductile adhesives inevitably exhibit nonlinear material behavior and can undergo inelastic or plastic deformation. Stresses and strains in the adhesive as well as in the adherends may be significantly affected by the presence of nonlinear material property.

Hart-Smith (1973a-d) has made an enormous contribution to this subject via the well-known PABST programme (Primary Adhesively Bonded Structure technology). He prepared four technical reports under the contract to the National Aeronautics and Aerospace Administration, USA. The elastic-plastic models shown in Figure 2.2(a) were used in his effort to include the plastic effect in the calculation of adhesive stresses. Hart-Smith developed analysis procedures and computer codes for four joint configurations, i.e. single lap, double lap, scarf and stepped, and doubler or lap shear joints.

There are also similar programs available from ESDU (Engineering Sciences Data Unit) for determining elastic and elastic-plastic stress and strain for double lap joints (ESDU, 1979) and stepped lap joints (ESDU, 1980). In the ESDU analysis procedures, the model used for characterizing adhesives is the ESDU/Grant model as given in equation (2.5) and shown in Figure 2.2(b).

These two analysis procedures represent a major and successful input to the analysis and design of lap joints. However, it should be borne in mind that both groups use the well-known shear lag model to calculate the non-linear shear strains and stresses in the adhesive layer. For most lap joints, the normal or peel stress acting across the bondline may be the main contributing factor to joint failure. Thus both peel and shear stresses should be considered simultaneously and should be solved together with the nonlinear adhesive characteristics. This proves to be quite challenging when analytical solutions are pursued.

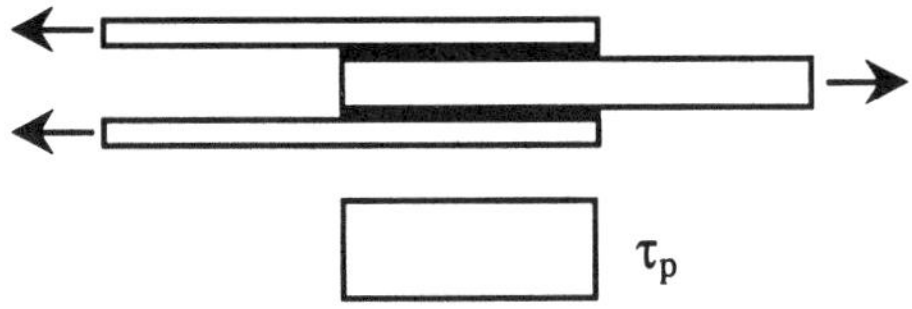

(a) Short overlap

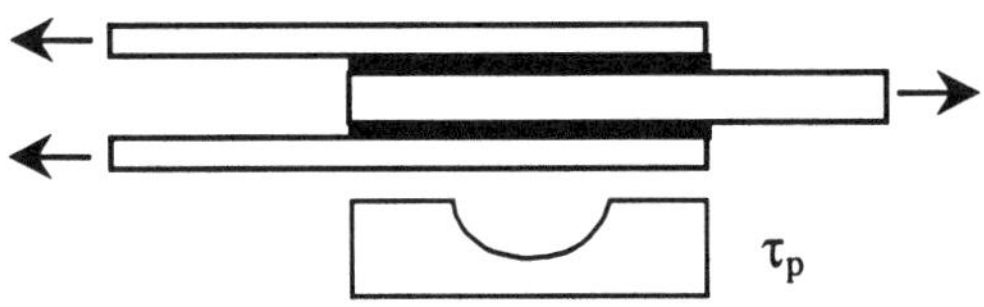

(b) Intermediate overlap

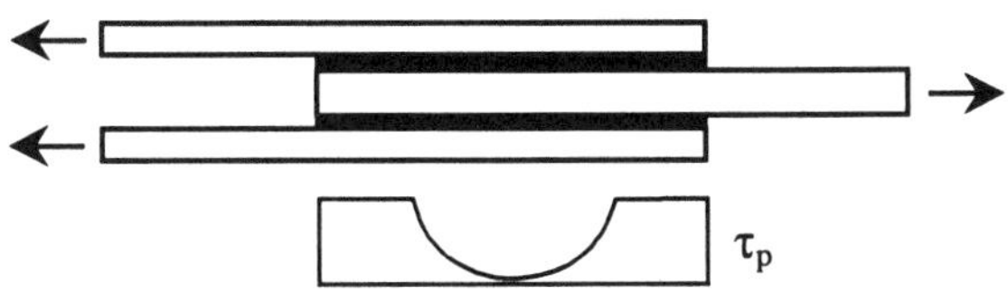

(c) Long overlap

Figure 2.11 Schematics of adhesive shear stress distribution with different overlap length

For balanced double lap joints with an elastic-plastic adhesive behavior as shown in Figure 2.2(a), Hart-Smith analysis (1973a) decouples the two governing equations for the peel and shear stresses based on the following grounds: (a) the shear stress near an overlap end becomes constant after yielding, and (b) the peak peel stress occurs at the overlap end. The nonlinear equation for adhesive shear stress is then solved and Figure 2.11 schematically depict the adhesive shear stress distribution for a double lap joint with a short, intermediate and long overlap. It is clear that at the overlap ends the shear stress is constant after yielding. When the joint overlap is short, the entire adhesive layer yields and a constant shear stress distribution is obtained. When the overlap length is intermediate, there exist an elastic trough in the overlap center and two plastic regions near the overlap ends. As the overlap length is increased from intermediate to long, the elastic trough enlarges and the minimum shear stress in the middle overlap approaches to zero. It is believed that existence of such elastic trough in a bonded joint is a major contributing factor to the improved fatigue life.

By assuming a linear elastic peel stress-strain characteristic, the governing equation for the peel stress can subsequently be solved. The explicit solution for the peak peel stress is

$$\frac{\sigma_{peak}}{\tau_{peak}} = \left(\frac{3E_c(1-\nu^2)t_o}{E\eta} \right)^{\frac{1}{4}} \tag{2.22}$$

where the peak peel stress σ_p is related to the peak shear stress τ_p (both at the end of the overlap). E_c and η are the effective transverse (peel) modulus and the thickness of the constrained adhesive layer. E, ν and t_o are the Young's modulus, the Poisson's ratio and the thickness of the outer adherend respectively.

Equation (2.22) clearly indicates that the peak peel stress is decreased when reducing the adherend thickness t_o and/or increasing the bondline thickness η. Thus it is recommended that overlap ends should be tapered in order to reduce the peak peel stress. There are in general two types of adherend-tapering schemes, as shown in Figure 2.12. The first adherend-tapering scheme has an adhesive layer of constant thickness, while the second one has an increased bondline thickness towards the adhesive end. These schemes have been proved to be an effective means of increasing lap joint strength. The adherend tapering can be achieved via machining for metallic adherends or dropping off plies for composite adherends.

The Hart-Smith plastic analysis of adhesive bonded joints demonstrates the importance of the nonlinear adhesive characteristics in the calculation of stresses and strains in adhesive joints. In the following subsection, we will describe an analysis procedure that takes the full non-linear adhesive characteristics into account.

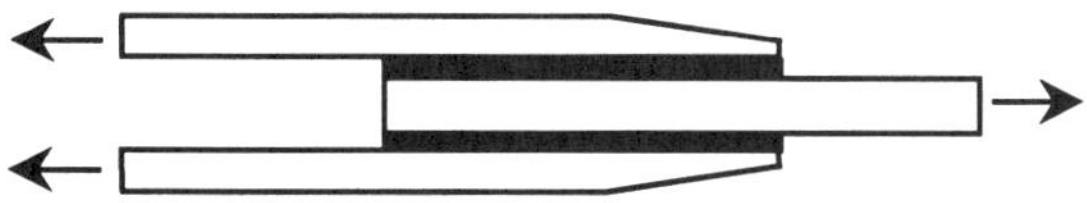

(a)

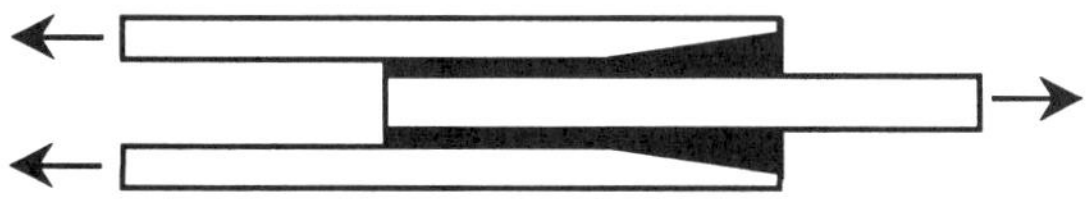

(b)

Figure 2.12 two typical tapering schemes at the overlap end: (a) unchanged bondline thickness, and (b) increased bondline thickness

(b) Sandwich model with nonlinear adhesive characteristics

The sandwich model, as shown in Figure 2.13, contains one or two adhesive layers and two or three adherends, and can be used to model typical joint overlap configurations. In this model, the adhesive layer is assumed to be subjected to only constant (across the bondline) shear and peel strains produced by differential straining of the two adherends. Similar to the spring-beam analogy (Cornell, 1953), the adhesive is assumed to exhibit nonlinear stress-strain characteristics in both shear and peel deformation, while the adherends are assumed to behave like a beam or cylindrically bent plate that only experience either longitudinal and lateral deflections or longitudinal deflection only.

<u>Model with two adhesive layers</u>

For a double lap joint, the adhesive sandwich model consists of an outer adherend, half of the inner adherend, and the adhesive with nonlinear adhesive properties. Due to symmetry, it is assumed that, while the outer adherend can experience longitudinal and lateral displacements, the inner adherend undergoes only longitudinal displacement (see Figure 2.14).

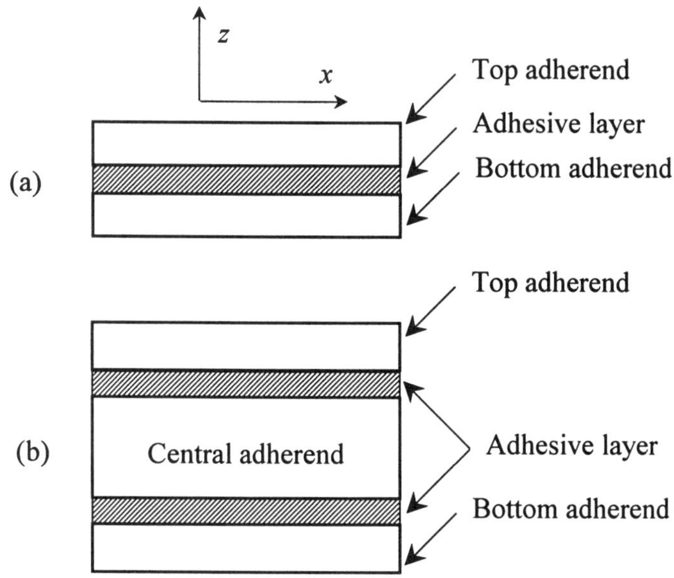

Figure 2.13 Adhesive adherend sandwich model containing one and two adhesive layers

The governing equilibrium, constitutive, and kinematic equations representing the adhesive-adherend model of Figure 2.14 can be written as follows:

Equilibrium equations:

$$N_{i,x} + 2\tau = 0 \qquad N_{o,x} - \tau = 0,$$
$$Q_{o,x} - \sigma = 0 \qquad M_{o,x} - Q_o + \frac{t_o}{2}\tau = 0 \tag{2.23}$$

Constitutive equations:

$$\begin{aligned} N_i &= A_i \varepsilon_i, \\ N_o &= A_o \varepsilon_o + B_o \kappa_o, \\ M_o &= B_o \varepsilon_o + D_o \kappa_o, \\ \sigma = \sigma_a(\varepsilon), &\qquad \tau = \tau_a(\gamma) \end{aligned} \tag{2.24}$$

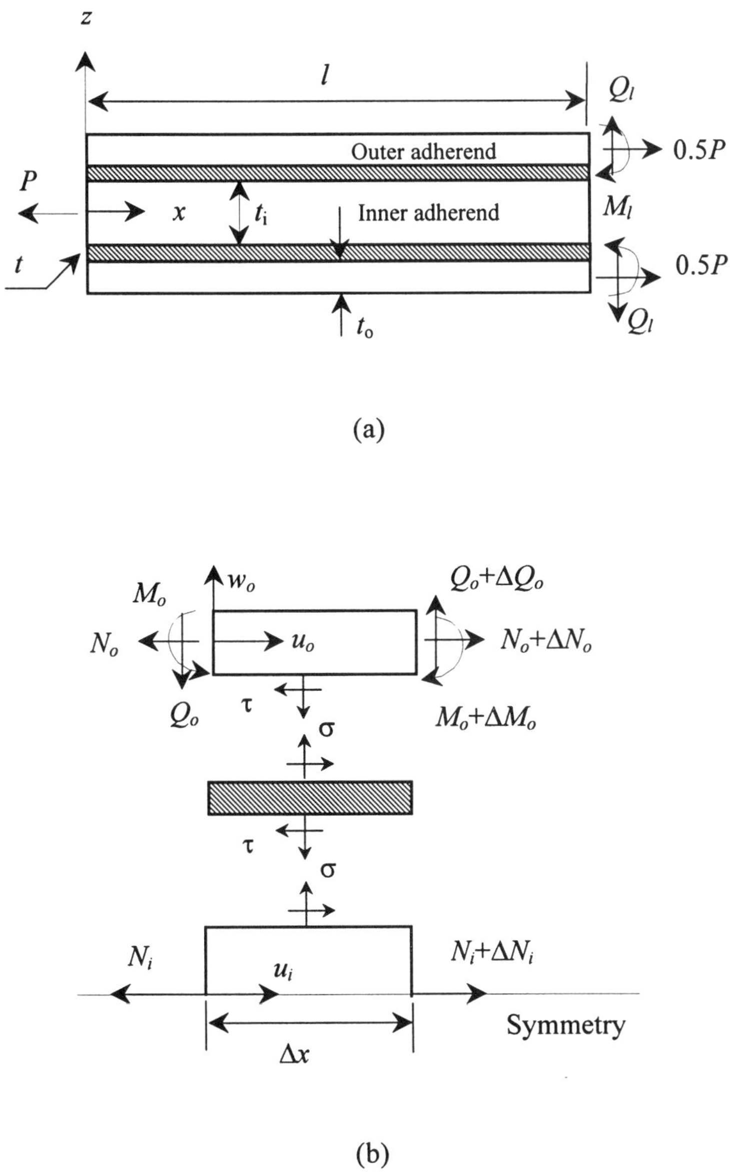

Figure 2.14 Adhesive adherend model with two adhesive layers: (a) an overlap with two adhesive layer, and (b) free body diagram of an infinitesimal element of the overlap

Kinematic equations:

$$\varepsilon_i = u_{i,x}, \qquad \varepsilon_o = u_{o,x}, \qquad \kappa_o = -w_{o,xx},$$
$$\varepsilon = \frac{w_o}{t}, \qquad \gamma = \frac{u_o - u_i}{t} + \frac{t_o}{2t} w_{o,x} \tag{2.25}$$

where the subscript i and o represent the inner and outer adherends, respectively, and N is the in-plane force resultant, M is the bending moment, Q is the transverse shear force, u is the mid-plane displacement, w is the out-of-plane displacement and t with a subscript is the thickness of the corresponding adherend and without a subscript is the thickness of adhesive layer. The parameters A and D are the axial and bending stiffness and the parameter B is the extension-bending coupling term. $\tau_a(\gamma)$ and $\sigma_a(\varepsilon)$ are the adhesive shear and peel stress-strain curves measured using thick adherend lap shear test or napkin ring shear test and bulk adhesive tensile test or butt joint tensile test, as described in Chapter 1.

To facilitate the numerical computation using the shooting method, we introduce the following variables:

$$y_1 = u_i, \qquad y_2 = u_o,$$
$$y_3 = N_i, \qquad y_4 = w_o,$$
$$y_5 = w_{o,x}, \qquad y_6 = Q_o, \qquad y_7 = M_o \tag{2.26}$$

Note that N_o is not included because it can be determined from the global equilibrium equation

$$N_i + 2N_o = P \tag{2.27}$$

Using the variables introduced in equation (2.26) and equation (2.27), we can rewrite equations (2.23) to (2.25) in the following form of seven first order differential equations (Tong and Jain, 1995):

$$y_{1,x} = y_3 / A_i,$$
$$y_{2,x} = A_o^* (P - y_3)/2 + B_o^* y_7,$$
$$y_{3,x} = -2\tau, \qquad y_{4,x} = y_5,$$
$$y_{5,x} = -B_o^* (P - y_3)/2 - D_o^* y_7,$$
$$y_{6,x} = \sigma, \qquad y_{7,x} = y_6 - \tau t_o / 2 \tag{2.28}$$

where A_o^*, B_o^* and D_o^* are given by

$$A_o^* = D_o / \Delta, \qquad B_o^* = -B_o / \Delta,$$
$$D_o^* = A_o / \Delta, \qquad \Delta = A_o D_o - B_o^2, \tag{2.29}$$

and the shear and peel strains in the adhesive layer can be expressed as:

$$\gamma = \frac{y_2 - y_1}{t} + \frac{t_o}{2t} y_5, \qquad \varepsilon = \frac{y_4}{t} \tag{2.30}$$

The peel and shear stress σ and τ in equation (2.28) are defined by the last two equations in equation (2.24) when there is no transverse stitching applied across the bondline or adherends. However, when it is assumed that the overlap is reinforced by transverse stitching through both adherends and the adhesive, contribution from the transverse stitching thread to the peel and shear stress σ and τ needs to be included (Tong and Jain, 1995).

Model with one adhesive layer

For a single lap joint, the adhesive sandwich model consists of two adherends and one adhesive layer with nonlinear adhesive properties (see Figure 2.15). It is assumed that both adherends undergo longitudinal and lateral displacements. Consider an infinitesimal element as shown in Figure 2.15(b). The governing equilibrium, constitutive and kinematic equations can be written as:

Equilibrium equations:

$$N_{1,x} + \tau = 0, \qquad N_{2,x} - \tau = 0,$$
$$Q_{1,x} + \sigma = 0, \qquad Q_{2,x} - \sigma = 0,$$
$$M_{1,x} - Q_1 + \frac{t_1}{2}\tau = 0, \quad M_{2,x} - Q_2 + \frac{t_2}{2}\tau = 0 \tag{2.31}$$

Constitutive equations:

$$N_1 = A_1\varepsilon_1 + B_1\kappa_1, \qquad M_1 = B_1\varepsilon_1 + D_1\kappa_1,$$
$$N_2 = A_2\varepsilon_2 + B_2\kappa_2, \qquad M_2 = B_2\varepsilon_2 + D_2\kappa_2,$$
$$\sigma = \sigma_a(\varepsilon), \qquad \tau = \tau_a(\gamma) \tag{2.32}$$

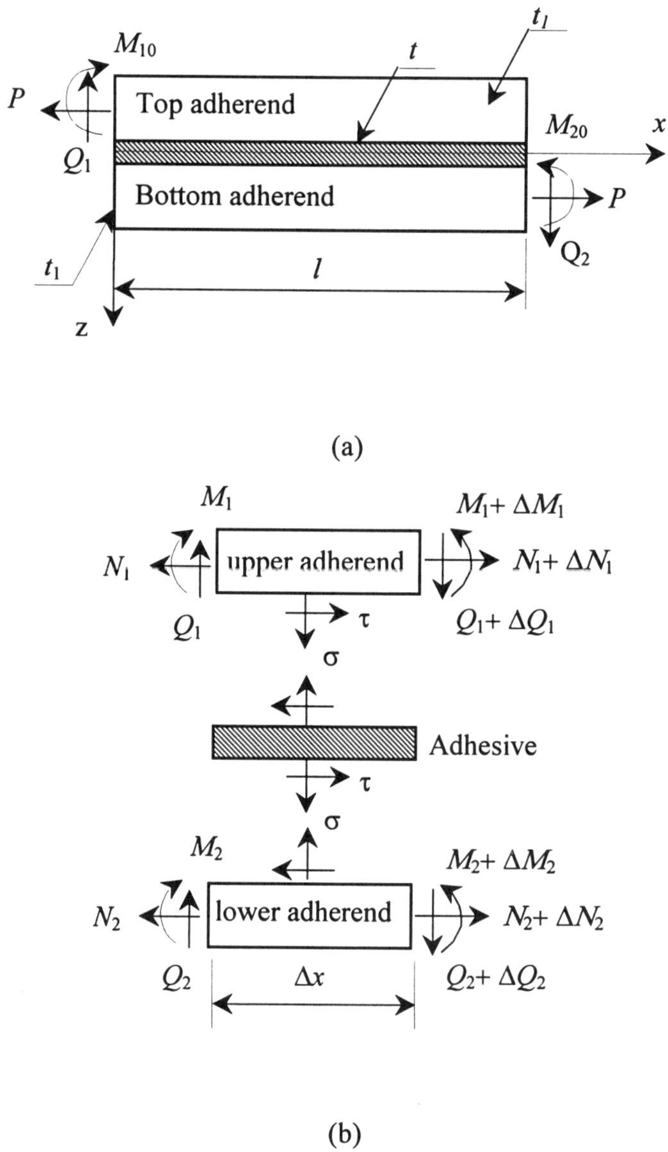

Figure 2.15 Adhesive adherend sandwich model with one adhesive layer: (a) an overlap with one adhesive layer, and (b) free body diagram of an infinitesimal element of the overlap

Kinematic equations:

$$\varepsilon_1 = u_{1,x}, \qquad \varepsilon_2 = u_{2,x},$$
$$\kappa_1 = -w_{1,xx}, \qquad \kappa_2 = -w_{2,xx},$$
$$\varepsilon = \frac{w_2 - w_1}{t}, \qquad \gamma = \frac{u_2 - u_1}{t} + \frac{1}{2t}(t_1 w_{1,x} + t_2 w_{2,x}) \tag{2.33}$$

where the subscript 1 and 2 represent the top and bottom adherends, respectively, and *A, B, D, N, M, Q, u, w* and *t* follow the definitions given in the preceding section.

Again, to simplify the equations and to facilitate the numerical computation, we define the following parameters

$$x = l\xi, \qquad \overline{u} = \frac{Pl}{A_1 + A_2}, \qquad \overline{w} = \frac{Pt}{E_a l} \tag{2.34}$$

and introduce the following variables y_i $(i=1,2,...,12)$

$$\begin{array}{lll} u_1 = y_1 & N_1 = y_2 & w_1 = y_3 \\ w_{1,x} = y_4 & Q_1 = y_5 & M_1 = y_6 \\ u_2 = y_7 & N_2 = y_8 & w_2 = y_9 \\ w_{2,x} = y_{10} & Q_2 = y_{11} & M_2 = y_{12} \end{array} \tag{2.35}$$

Using the above variables, we can rewrite equation (2.31) to (2.33) in the form of twelve first order differential equations:

$$\begin{array}{ll} y_{1,x} = A_1^* y_2 + B_1^* y_6, & y_{2,x} = -\tau \\ y_{3,x} = y_4, & y_{4,x} = -B_1^* y_2 - D_1^* y_6, \\ y_{5,x} = -\sigma, & y_{6,x} = y_5 - \dfrac{t_1}{2}\tau \\ y_{7,x} = A_2^* y_8 + B_2^* y_{12}, & y_{8,x} = \tau \\ y_{9,x} = y_{10}, & y_{10,x} = -B_2^* y_8 - D_2^* y_{12}, \\ y_{11,x} = \sigma, & y_{12,x} = y_{11} - \dfrac{t_2}{2}\tau \end{array} \tag{2.36}$$

where the constants A_i^*, B_i^* and D_i^* (i=1,2) are defined similar to the expressions given in equation (2.29).

The shear and peel strains in the adhesive layer can be expressed as:

$$\varepsilon = \frac{1}{t}(y_9 - y_3)$$

$$\gamma = \frac{1}{t}(y_7 - y_1) + \frac{1}{2t}(t_1 y_4 + t_2 y_{10}) \tag{2.37}$$

The nonlinear equations in equation (2.28) and (2.36) can be solved numerically using various numerical schemes, such as finite difference method, shooting method etc (Press et al 1992). The solutions to these two sets of equations can provide appropriate information for analysis and design of bonded lap joints according to the comparison of various lap joint theories (Carpenter, 1991). The two sets of equations were solved using the shooting method Tong and his colleagues (Tong and Jain, 1995; Tong, 1997; Tong, Kurrupu and Kelly, 1997), and a software package, called DAJOINT, was developed with a user-friendly graphic interface. The software program can be used to analyze stresses in adhesive bonded single lap, double lap and lap shear joints with and without considering the transverse stitching effect (Tong and Jain, 1995). Readers who are interested in the software program are recommended to contact the authors.

It should be pointed out that equations similar to equation (2.28) and (2.36) have been solved by other researchers, who considered either linear or particular nonlinear material characteristics. For example, Yuceoglu and Updike (1980) presented a linear stress analysis for bonded plates and joints. Picket and Hollaway (1985) presented an analysis of elastic-plastic adhesive stress in lap joints. Roberts (1989) assumed a linear adhesive behavior and developed a two-stage analytical procedure for determining the distribution of shear and peel stresses in a variety of adhesive joints. Bigwood and Crocombe (1989) and Crocombe et al (1990) presented an elastic analysis and engineering design formulae for bonded joints.

(c) Illustrative examples

In this subsection, some examples will be presented to demonstrate the effect of the nonlinear adhesive characteristic, overlap length, adherend thickness and the end tapering. For simplicity, only the results for the balanced double lap joint configuration will be presented.

The nonlinear adhesive characteristic, as shown in Figure 2.1, is selected to demonstrate the effect of nonlinear adhesive stress-strain curves and the overlap length on the stress distribution in the adhesive layer. The thickness of the outer and inner adherend is 1 and 2 mm, and the overlap length is chosen to be 10, 20 and 30 mm respectively. The Young's modulus and Poisson's ratio of the adherend are 70 GPa and 0.33. For the linear case, the adhesive has a Young's modulus of 2.518 GPa and a shear modulus of 1 GPa (as can be seen from Figure 2.1). The applied load is chosen to be 950 N/mm. Figures 2.16 and 2.17 depict the shear and peel stress distributions in the adhesive layer for the balanced double lap joints with nonlinear and linear adhesive characteristics, respectively.

As shown in Figure 2.16 and 2.17, the predicted peak adhesive stresses for the case of a non-linear adhesive characteristic are much lower than those predicted for the linear adhesive case. For example, the peak peel stress is only 38 MPa for the nonlinear adhesive behavior and is as large as 79 MPa for the linear adhesive behavior. In addition, there also exists a significant difference in the shear stress distribution patterns for both cases, especially for the joint with a relatively short overlap. It is noted that the overlap length has a significant effect primarily on the shear stress distribution. Similar to the result in Figure 2.11 (Hart-Smith, 1973b), the whole adhesive layer undergoes plastic deformation when the joint overlap is short, there exists an elastic trough in the middle of the overlap and two plastic deformation zones at the two ends for intermediate and long joint overlaps.

The double lap joint shown in Figure 2.18(a) has a Young's modulus of 70 GPa and a Poisson's ratio of 0.3. The overlap length is 20 mm, the thickness of the inner adherend and outer adherend is 3 and 1.5 mm respectively. The end thickness is tapered down to 0.5 mm with a thickness of 1 mm being removed, and the tapering ratio varies from 5:1 to 15:1 with an increment of 2.5. To demonstrate the effect of end tapering, we introduce a quantitative measure called relative difference, which is defined as (a) $(\sigma_{tp} - \sigma_{utp})/\sigma_{utp}$ for the peel stress, and (b) $(\tau_{tp} - \tau_{utp})/\tau_{utp}$ for the shear stress. The subscripts *tp* and *utp* stand for the tapered and non-tapered joints respectively.

Figure 2.18(b) plots the relative difference in percentages in the adhesive peak stresses versus the tapering ratio. Clearly, higher the tapering ratio, the more significant the reduction in the peak adhesive stresses. However, the reduction in the relative difference percentage seems to approach limit values when the tapering ratio becomes larger than 10.

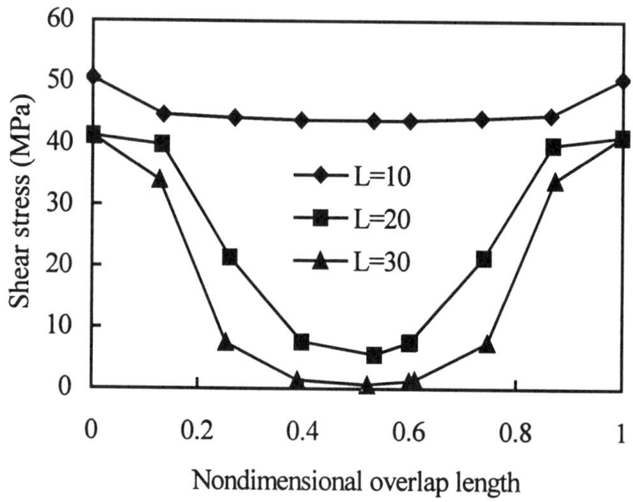

(a) Shear stress distribution

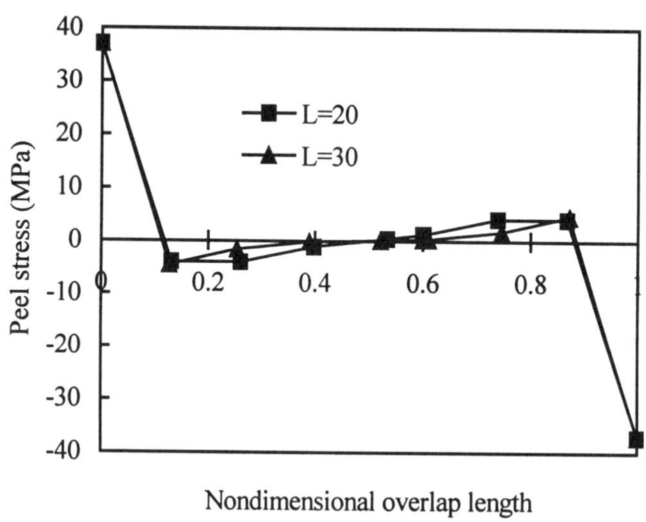

(b) Peel Stress distribution

Figure 2.16 Shear and peel stress distribution in the adhesive layer for adhesive bonded double lap joints with different overlap length of *L* with a nonlinear adhesive characteristic

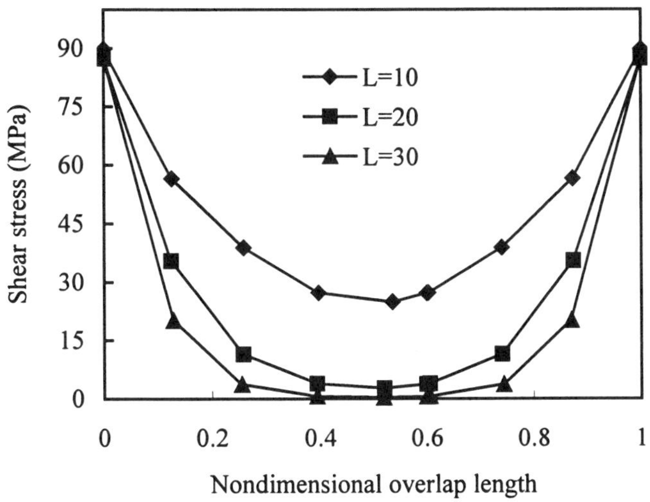

(a) Shear stress distribution

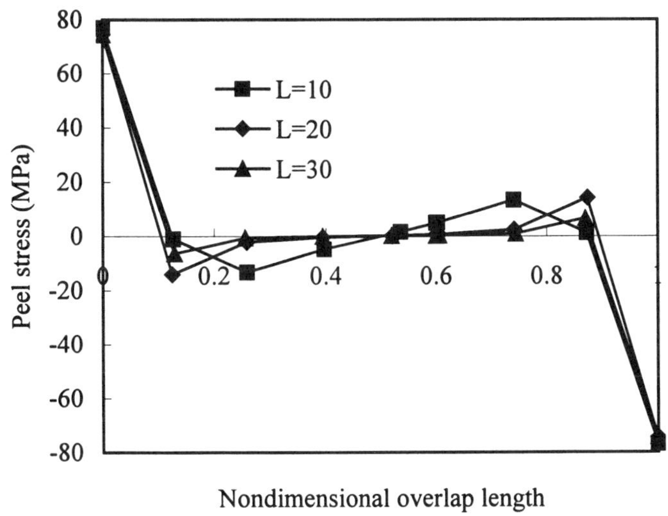

(b) Peel Stress distribution

Figure 2.17 Shear and peel stress distribution in the adhesive layer for adhesive bonded double lap joints with different overlap length of *L* with a linear adhesive characteristic

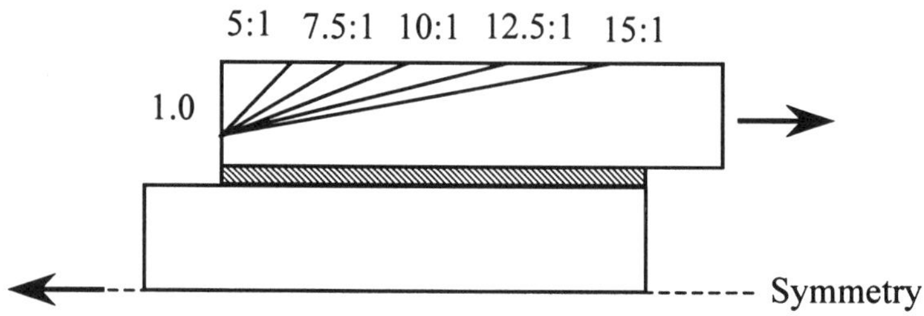

(a)

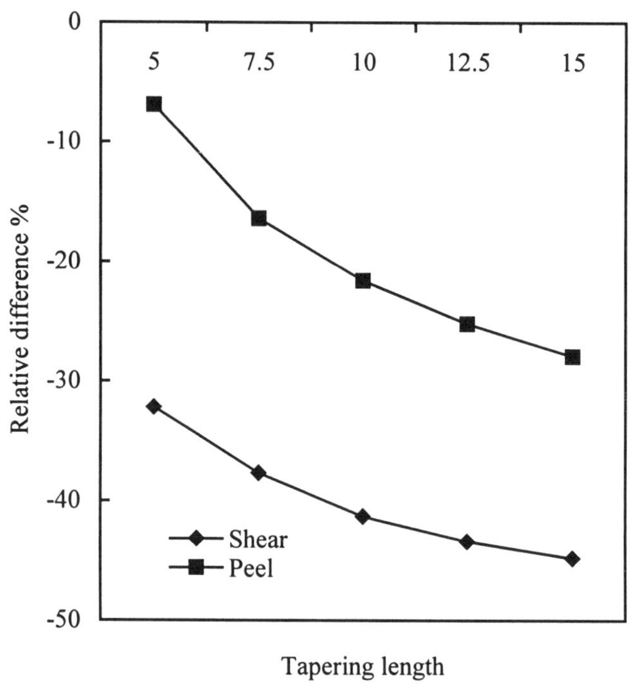

(b)

Figure 2.18 Effect of tapering ratio, i.e., tapering length/tapered thickness

The double lap joint of Figure 2.19(a) is considered to study the effect of adherend thickness on the peak adhesive stresses. The adherend properties are E=70 GPa and ν=0.3. The overlap length is taken to be 20 mm. The double lap joint is assumed to be a balanced one, i.e. the thickness of the outer adherend is assumed to be half of that of the inner adherend. The tapering ratio is kept as a constant of 1:10, and the thickness left after tapering at the outer adherend end is taken to be 0.5 mm. The outer adherend thickness is then varied from 1 mm to 2.5 mm with an increment of 0.25 mm, which in turn results in an inner adherend thickness variation from 2 to 5 mm with an increment of 0.5 mm. The relative differences are defined as above for both peak peel and shear stresses and are plotted versus the outer adherend thickness in Figure 2.19(b). It is clearly shown that end tapering tends to yield a more significant reduction in the peak adhesive stresses for balanced double lap joints with thicker adherend when both the tapering ratio and the thickness left remain unchanged. For example, reduction in the peak peel stress is only about 8% when the outer adherend is 1 mm thick and increases up to about 25% when the outer adherend is 1.5 mm thick. It is also worth pointing out that the tapering length is only 5 mm for an outer adherend of 1 mm thick, and becomes 20 mm for an outer adherend of 2.5 mm thick.

2.3.4 Two dimensional analysis

In the Goland and Reissner's type of analysis, the longitudinal stress in the adherend varies linearly throughout the adherend thickness to facilitate bending and stretching action, and the stresses in the adhesive are constants. However, variations in the shear and normal stresses throughout the thickness of the adherend and adhesive layer are not taken into account. As a result, the exact boundary condition of zero shear stress at the ends of the square edged adhesive layer can not be satisfied. Thus a research effort has been made towards overcoming these two deficiencies in the past two decades, for example, Allman (1977), Chen and Cheng (1983) Adams and Mallick (1992). In their analysis, the two two-dimensional elasticity equilibrium equations are used to express the shear and normal stresses in terms of the through-thickness coordinate y, i.e.

$$\frac{\partial \sigma_x}{\partial x} + \frac{\partial \tau_{xy}}{\partial y} = 0, \quad \frac{\partial \tau_{xy}}{\partial x} + \frac{\partial \sigma_y}{\partial y} = 0$$

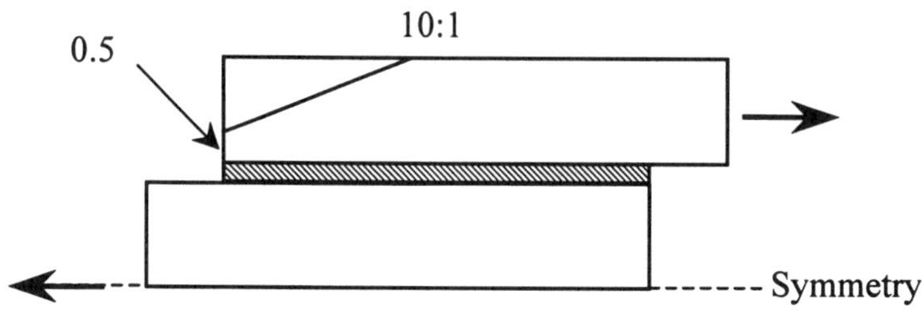

(a)

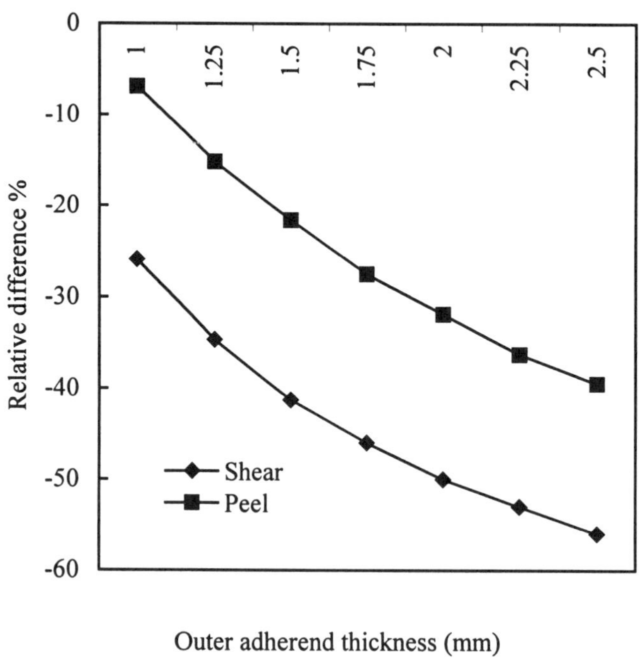

(b)

Figure 2.19 Effect of end tapering on peak shear and peel stresses in the adhesive layer of the double lap joint with a variety of adherend thickness

In this book the analysis based on the above two equations is referred to as two-dimensional analysis while the Goland and Reissner type analysis is referred to as one-dimensional analysis.

In the two-dimensional analysis, the longitudinal stress σ_x is assumed to vary linearly throughout the adherend thickness. By integrating the above two equations with respect to the coordinate y, the expressions for the shear and normal stresses τ_{xy} and σ_y can be then determined. For the adhesive layer, the shear stress is assumed to be constant across the adhesive thickness, and integration of the above 2nd equation results in an expression of the normal stress varying linearly across the adhesive thickness. After formulating expressions for the stresses in both adherend and adhesive layer, the total complementary strain energy for the joint structure is then minimized to impose the compatibility conditions, which in turn can be solved to determine the stress distributions. It has been shown that the stresses determined following this procedure satisfy the zero shear stress boundary condition at the ends of the squared-edged adhesive layer.

In addition, there are some models that only consider variations of the shear and normal stresses in either adherend or adhesive. Renton and Vinson (1977) took into account the variations of stresses through adherend thickness by using a mixed displacement and stress function formulation with constant stresses in adhesive layers. Ojalvo and Eidinoff (1978) assumed that the shear stress in the adhesive layer varies linearly and the peel stress remains constant, and employed the classical beam theory to model the adherends. Their purpose was to investigate the effect of the bondline thickness on stress distribution and failure location.

It is noted that while the two-dimensional analysis satisfies the zero shear stress boundary conditions from an academic point of view it becomes more complex than the one-dimensional analysis. Another question is the existence of a square-edged adhesive end in practice as adhesive in liquid state overflows and forms a spew fillet at an overlap end. Although the shear stress along the spew fillet surface should be zero, this is different from the zero shear stress condition at the square-edged adhesive end.

2.4 FINITE ELEMENT ANALYSIS

The finite element method is one of the most powerful, robust and versatile numerical techniques that has been widely used in almost all fields involving engineering computation. There exist a large number of texts describing various aspects of the finite element analysis (Zienkiewicz and Taylor, 1989). When the finite element method is used in the structural analysis, the responses of a structure in an arbitrary shape subject to a variety of boundary conditions and loading environments can be evaluated. Wooley and Carver (1971) and Adams and Peppiatt (1974) are amongst those who first employ the finite element method to determine the stress distributions in adhesive bonded joints. Nowadays, the finite element method has been widely used in analysis and design of adhesive bonded joints. A variety of important aspects, such as complex geometry, nonlinear material property and large deformation, can be taken into account in the analysis due to fast advance of computing power, such as computer speed and size of memory. However, it is not an easy job to establish an appropriate finite element model for an adhesive bonded joint. There are a number of questions that need to be carefully considered before an adequate finite element model can be developed for analyzing an adhesive bonded joint. For instances, What type of model should be used to model the joint geometry? Should the nonlinear material behavior be modeled? If so, what type of material model should be used? It is not an easy task to address these questions. In this section, an attempt is made to discuss these questions from a number of aspects of modeling strategy.

In the finite element stress analysis of an adhesive bonded joint, an engineer needs to consider three main aspects: description of the geometry, modeling of the materials, and simulation of the loading environment. They should be examined separately and then carefully integrated in order to develop a simple and efficient finite element analysis solution for an adhesive bonded joint.

2.4.1 Meshing schemes

Definition of the geometry of an adhesive bonded joint includes both adherends and adhesive layer. It is relatively easier to define the geometry of adherends, even in complex shapes, as per the design. However, definition of the geometry of adhesive layer is not a trivial job. Near the end of an overlap, an adhesive spew fillet can often be formed due to adhesive overflow in the adhesive consolidation procedure. The shape and size of the adhesive spew fillet depends on many manufacturing parameters, such as pressure, curing temperature and the amount of adhesive applied between the two adherends. In fact, the size and shape of the spew fillet are manufacturing related parameters, which possess irregularity and uncertainty and can only be measured after making joint. In all the finite element analyses, there are in general two types of models, i.e. one that does not model adhesive spew fillet as shown in Figure 2.20(a) and another one that models the spew fillet as schematically shown in Figure 2.20(b)-(d).

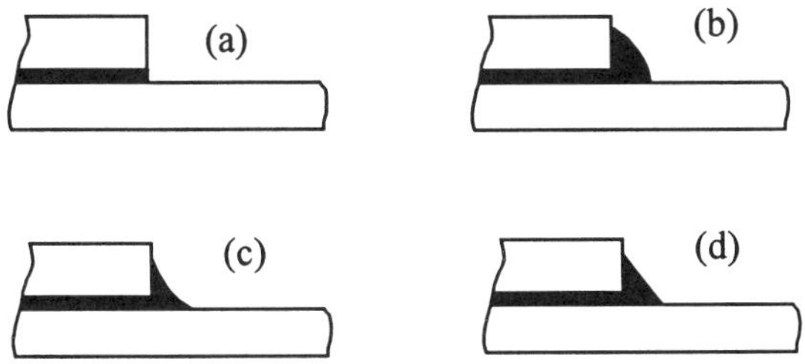

Figure 2.20 Schematics of real and ideal adhesive spew fillets

In addition, voids or defects due to entrapped air bubbles often exist in the spew fillet or one the surface of spew fillet, which makes modeling of the spew fillet more complicated. These voids can grow when adhesive is subject to loading and coalesce into a crack and further propagate to cause premature failure. The failure mechanisms and strength prediction will be described in Chapter 3.

Although it is desirable to model the geometry of a joint with as much detail as possible there always exists a limit where there needs a compromise between cost and accuracy. In reality, every joint is a three-dimensional (3D) structure. For example, the adhesive bonded doubler specimen as shown in Figure 2.21a modeling a portion of a stiffened composite plate (Kassapoglou, 1993, 1994; Tong, Sheppard and Kelly, 1994). Both adherends and adhesive layer can be modeled using 3D brick elements, such as 8-node and 20-node brick elements and 6-node wedge elements. As high stress gradients are expected a fine mesh should be used near the overlap end and along the edges in the width direction. In the through-the-thickness direction, a ply-by-ply meshing scheme may be used to model each individual ply and adhesive layer, namely, one element for a ply or the adhesive layer. For example, if 20 elements in the length direction and 8 elements in the width direction are used for modeling the overlap and the unsupported part of the adherend respectively, 4000 brick elements in total will then be required for such a simple joint with both adherends having 8 plies. While it can be handled effectively for linear analysis, it becomes quite a big job when large deformation and nonlinear material properties are taken into account in the analysis. In addition, if a finer mesh is required, say for example with two or more elements are used in the through-the-thickness direction, obtaining a full 3-D finite element solution becomes very expensive.

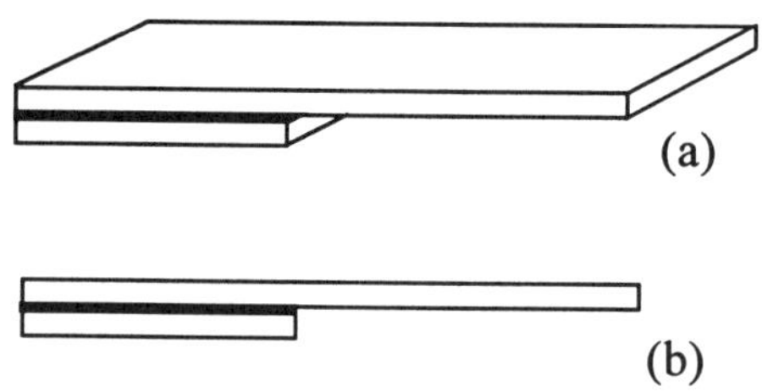

Figure 2.21 Adhesive bonded doubler representing skin-stiffener portion of a stiffened plate. (a) a full 3D view, (b) a side view

One question that can be asked is when a two-dimensional (2-D) plane-strain or plane-stress analysis should be used instead of a full 3D analysis. Richardson et al (1993) presented a comparison between the two- and three-dimensional finite element analysis of adhesive joints. Their results show that for a number of bonded joints a 2-D finite element solution is more than adequate. For the side view cross section of the doubler shown in Figure 2.21(b), a 2-D plane-strain finite element analysis only requires 500 quadrilateral elements using the same meshing scheme as a full 3-D analysis. Another difference is that an 8-node brick element uses a nine-point (3 by 3) Gaussian integration quadrature while a 4-node quadrilateral element employs a four-point (2 by 2) Gaussian integration quadrature when calculating the element stiffness matrix (Zienkiewicz and Taylor, 1989). Figure 2.22(a) depicts a 2-D plane strain model for the doubler using one element to model adhesive layer and eight elements to model both adherends in the thickness direction. A finer mesh can also be developed by using more elements in the thickness direction in the adhesive layer. In this instance, the mesh in the adherends is also needed to be refined as there exists a limit on the aspect ratio, for example, 1 in 100, for a quad element in MSC/Nastran (MacNeal-Schwendler, 1994).

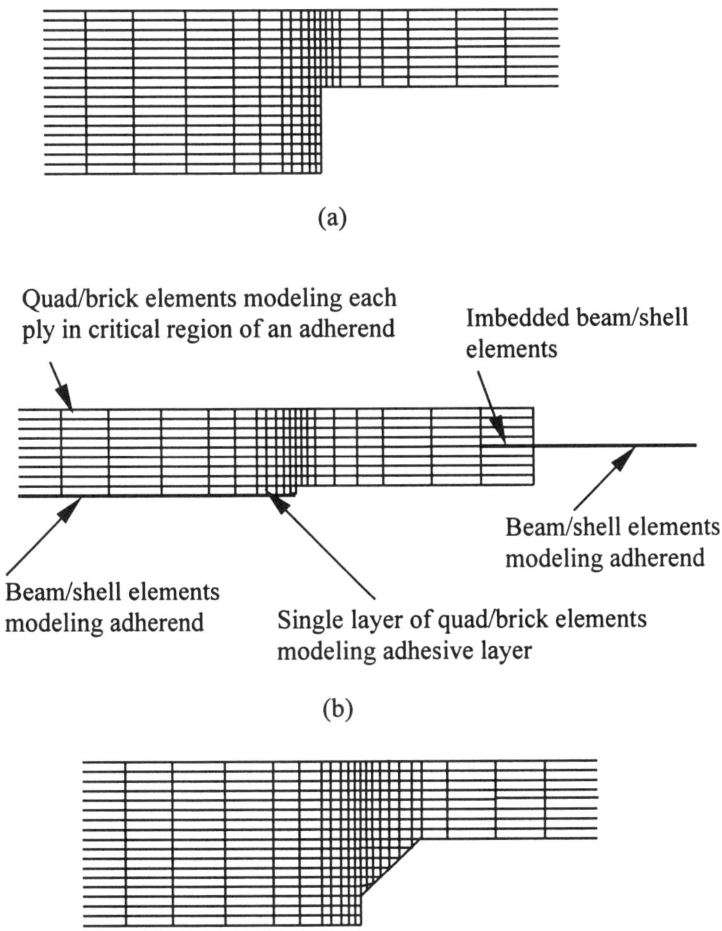

(c)

Figure 2.22 Modeling schemes for an adhesive bonded doubler. (a) a 2-D plane strain model, (b) a mixed beam/shell and quad/brick element model, and (c) a 2-D plane strain model with spew fillet

The doubler can also be modeled in a mixed format as shown in Figure 2.22(b). In the case of a 2-D mixed model, beam and quadrilateral elements can be used to model the adherend, while in the case of a 3-D mixed model, shell and brick elements can be used together to save computation effort. To connect beam/shell elements to quad/brick elements, one can imbed one beam/shell element with a fraction of membrane stiffness and a full bending stiffness or impose constraints at the joining cross section. When beam/shell elements are connected to the adhesive quad/brick elements by using an offset neutral plane facility in most FEA code, for example, MSC/Nastran (MacNeal-Schwendler, 1994).

Another 3-D finite element modeling scheme is to model both plates/adherends using plate/shell elements and model the adhesive layer using spring elements, which transfer normal force and two perpendicular shear forces. The 2-D equivalence is to model both adherends with beam elements and the adhesive layer with spring elements. These modeling schemes can be developed using the analytical governing equations in conjunction with the Ritz method or virtual work principle.

For some joint configurations, the adherends come in complex shape. Tapering of an adherend needs to be modeled properly as it can significantly affect the stress distribution. When the shape of the adhesive spew fillet is known, the corresponding finite element mesh can be generated although it can be very complicated. For example, Figure 2.22(c) shows an example of meshing the triangular spew fillet.

In addition to an adhesive spew fillet and complex adherend geometry, there may exist a delamination crack in the composite laminate or void in adhesive. Modeling of such detailed geometry can significantly complicate mesh generation. The mesh around a crack in a composite laminate will be discussed in Chapter 4 to address damage tolerance issue of bonded joints.

For tubular or but type of joints, a simple axi-symmetric meshing scheme may be used when axi-symmetry of geometry and loading is assured. It is worth noting that such condition may be violated due to manufacturing tolerance. For example, an ideal butt joint in tension is axi-symmetric, but eccentricity may exist if great care is not taken in preparing the specimen. Such eccentricity can greatly affect the stress distribution and further complicate the failure stress determination (e.g. Adams and Coppendale, 1977).

2.4.2 Material properties

When both adherends and adhesive can be assumed to be linear and then only the Young's modulus and Poisson ratio are required to model the material behavior. However, when subjected to hostile operating environment such as a high load and an elevated temperature, the adhesive may exhibit a non-linear material behavior, such as an elastic-plastic behavior. In addition, some adherends may also exhibit nonlinear material behaviors, for example, an aluminum adherend in peel test. Elastic-plastic analyses of various types of joints have been conducted by many researchers, for example, Crocombe and Adams (1981, 1982) presented an elastic-plastic investigation of the peel test, Picket and Hollaway (1985) presented an analysis of elastic-plastic adhesive stress in bonded lap joints, Edlund and Klarbring (1993) and Edlund (1994) presented a coupled elastic-plastic damage model for rubber-modified epoxy adhesive.

The adhesives, particularly the ductile adhesives, often exhibit a nonlinear elastic-plastic behavior as described in Section 2.2. Adhesive yielding can be determined using various yield criterion, which define the condition that characterizes the change of material from an elastic state to a plastic state. When a material is subject to a uniaxial stress, a yield criterion is defined as, when the stress attains the yield stress of the material; whereas when subject to multiaxial stresses, a yield criterion defines which combination of the multiaxial stresses will cause yielding. For an isotropic material the yield criterion must be a symmetric functions of the principal stresses σ_1, σ_2 and σ_3:

$$f(\sigma_1, \sigma_2, \sigma_3) = \text{constant}$$

Presented below are some widely used yield criteria which have been implemented in most commercially available finite element analysis software, e.g. MSC/Nastran (MacNeal-Schwendler, 1994).

Tresca yield criterion (1864): yielding occurs when the maximum tangential stress attains $0.5\sigma_s$, (where σ_s is the tensile yield stress), i.e. when one of the following three conditions is true

$$\begin{aligned} 2|\tau_1| &= |\sigma_2 - \sigma_3| = \sigma_s \\ 2|\tau_2| &= |\sigma_3 - \sigma_1| = \sigma_s \\ 2|\tau_3| &= |\sigma_1 - \sigma_2| = \sigma_s \end{aligned} \tag{2.38}$$

The above conditions define a right hexagonal cylinder with axis $\sigma_1=\sigma_2=\sigma_3$ perpendicular to the deviatoric plane. When $\sigma_1 > \sigma_2 > \sigma_3$, the Tresca criterion can be simplified as follows: yielding occurs when

$$|\tau_{\max}| = 0.5\sigma_s$$

Von Mises yield criterion (1913): yielding occurs when the following condition becomes true

$$(\sigma_1 - \sigma_2)^2 + (\sigma_2 - \sigma_3)^2 + (\sigma_3 - \sigma_1)^2 = 2\sigma_s^2 \tag{2.39}$$

This criterion can also be described as yielding begins when the elastic energy of distortion exceeds a critical value.

Mohr-Coulomb yield criterion (1773): yielding occurs when

$$\tau = c - \sigma_n \tan\phi \tag{2.40}$$

where τ is the magnitude of the shearing stress, σ_n is the normal stress (tensile stress is positive), c is the cohesion and ϕ is the angle of internal friction. When $\sigma_1 > \sigma_2 > \sigma_3$, the above equation becomes

$$\sigma_1 - \sigma_3 = 2c\cos\phi - (\sigma_1 + \sigma_3)\sin\phi$$

Drucker-Prager yield criterion (1953): Drucker and Prager modified the Von Mises criterion by including the influence of a hydrostatic stress component on yielding, namely,

$$\frac{2\sin\phi}{\sqrt{3}(3-\sin\phi)}(\sigma_1 + \sigma_2 + \sigma_3) + \sqrt{(\sigma_1\sigma_2 + \sigma_2\sigma_3 + \sigma_3\sigma_1)} = \frac{6c\cos\phi}{\sqrt{3}(3-\sin\phi)} \tag{2.41}$$

Once plastic deformation takes place in the material, strain hardening or plastic flow rules need to be applied. For example, one can use isotropic hardening rule, or kinematic hardening rule or even a combination of the isotropic and kinematic rules. Keen readers are recommended to read the relevant texts on plasticity, e.g. the text by Owen and Hinton (1980).

2.4.3 Solution techniques

Linear analysis method can be used for joints experiencing small deformation and having linear material properties. There are a variety of joints that experience a large deformation and/or large rotation. For example, a single lap joint has a large deformation and rotation that must be taken into account to obtain correct stress and strain values. When there is a large deflection or rotation or nonlinear material behavior, an iterative nonlinear solution procedure must be employed to achieve converged solutions for stresses and strains in joints at each incremental load step.

When a detailed stress information is required, a global and local analysis scheme may be employed to conduct a global analysis first followed by a local analysis. For example, a global analysis of an adhesive bonded stiffened plate can be carried out using plate elements with a coarse mesh used in the stiffener area or even model the stiffener using beam element. A detailed finite element model can then be used for the plate and stiffener interface to investigate the interfacial debonding behavior in conjunction with the deformations obtained in the global analysis.

2.4.4 Some illustrative results

There exist a large amount of information on stresses and deformations in various adhesive bonded joints. In this subsection, we intend to address some important issues of the numerical results rather than to review the results available. The chosen issues are mesh dependency, spew fillet effect, large deformation and/or rotation, and nonlinear material behavior.

(a) Mesh dependency

One of the major characteristics of adhesive bonded joints is the existence of bimaterial interfaces between adherends and adhesive layer. Stress concentrations can occur at the end of a bimaterial interface. Many researchers, e.g. Hein and Erdogan (1971), have addressed such types of stress concentration problems. At the end of an adhesive layer, stress singularities can exist at the end of the adherend and adhesive interface (Delale, 1984; Groth, 1988, Hattori, 1991).

To demonstrate the stress concentration at the adhesive interface end, let us consider a double lap joint in tension as shown in Figure 2.23a. TheYoung's modulus and Poisson ratio are taken to be 70 GPa and 0.3 for adherend and 2.5 GPa and 0.35 for the adhesive. As shown in Figure 2.23a, the applied stress is 100 MPa.

A 2D plane strain finite element model is created for the joint using 4-node quadrilateral elements. The initial mesh is generated by subdividing the outer adherend, the adhesive and the inner adherend with 10 by 20, 1 by 20 and 10 by 40 divisions, which yields the same element size of 0.2 by

1 mm. Based on this initial mesh, three finer meshes are generated by further subdividing the elements locating between x=18 mm to 22 mm. The first mesh is generated by subdividing the elements within this range with a 1 by 5 division, and the smallest square element size near the interface end is 0.2 mm. The second mesh is obtained by further subdividing these square elements of 0.2 mm long with a 2 by 2 division yielding the smallest square element of 0.1 mm long. The third mesh is obtained by further subdividing the second mesh with a 2 by 2 division, which yields the smallest elements of 0.05 mm long. Three linear finite element solutions are obtained using Strand6 (G+D, 1995). Figures 2.24 and 2.25 plot the shear and peel stress distributions near the overlap end along the centerline of the adhesive and the interface between the adhesive and inner adherend.

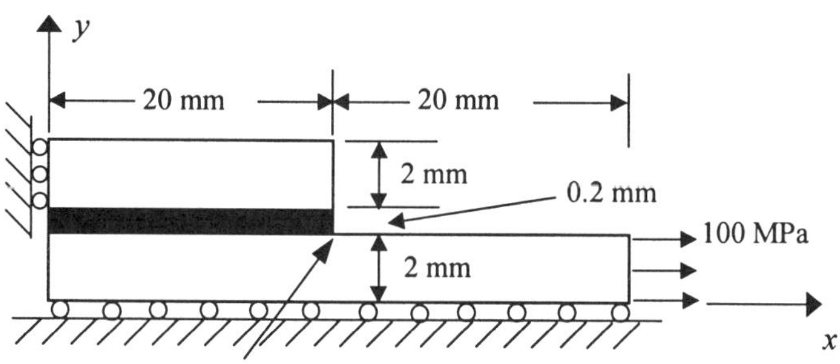

Figure 2.23a Double lap joint in tension

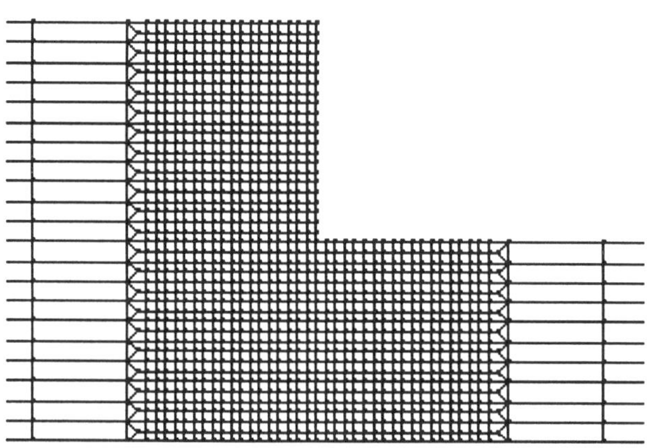

Figure 2.23b Mesh details near overlap end in the double lap joint

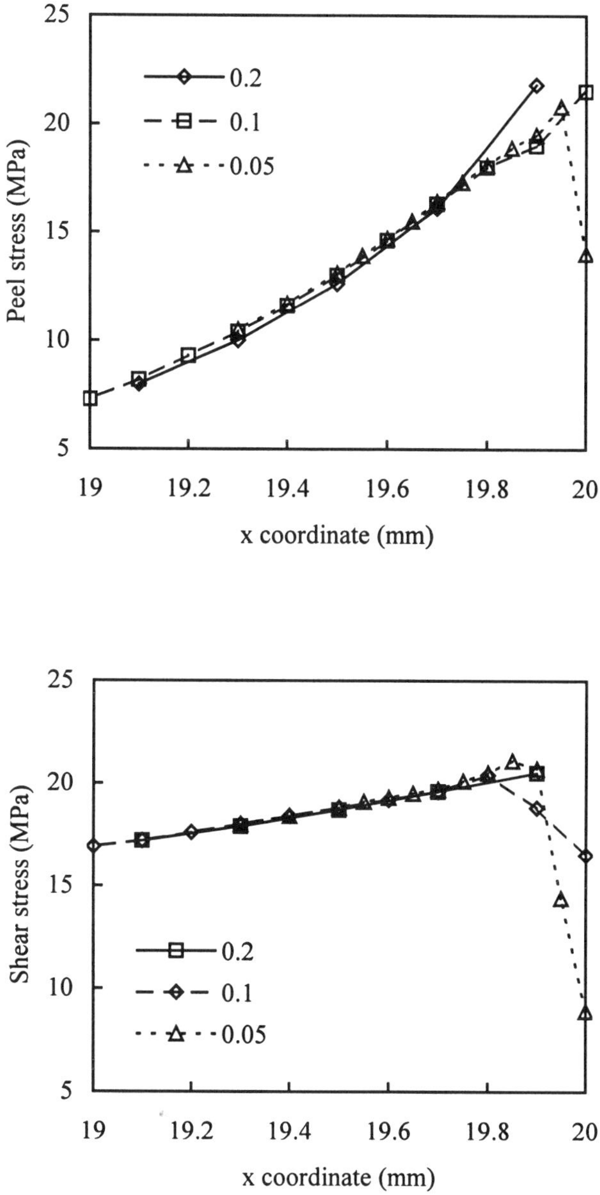

Figure 2.24 Peel and shear stress distribution along the centerline of the adhesive layer (y=2.1 mm)

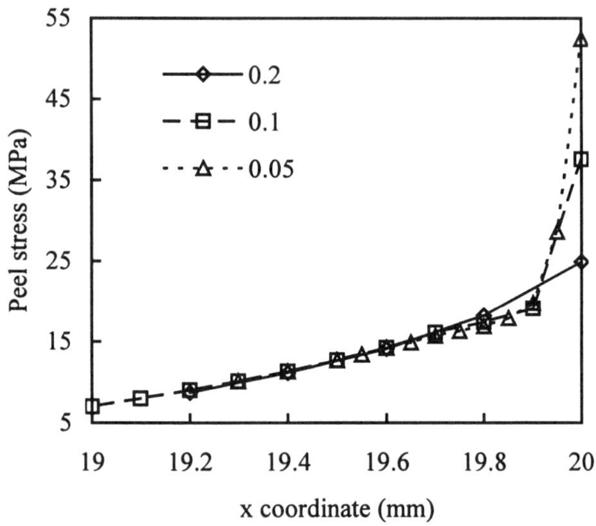

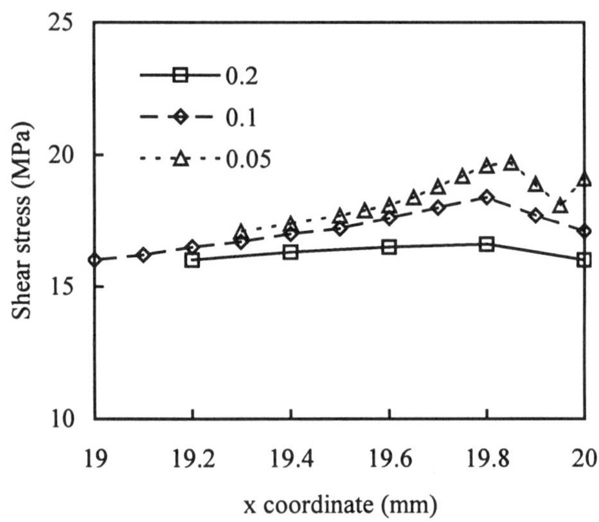

Figure 2.25 Effect of element size on the peel and shear stress distribution along the interface between the adhesive and the inner adherend (y=2 mm)

Figure 2.25 shows that the peel stress at the adhesive interface end tends to increase when the element size is decreased from 0.2 to 0.05 mm. This clearly shows that the peel stress at the adhesive interface end does not converge as the element size is reduced, and thus indicates the peel stress is mesh dependent. The mesh dependency can create difficulty in predicting joint strength using point-based maximum stress or strain criteria. This important issue will be discussed in Chapter 3.

(b) Spew fillet effect

The spew fillet is referred to as the excess portion of adhesive squeezed out from the overlap area. The effect of spew fillet on stress distributions has been studied by Adams and his colleagues (Crocombe and Adams, 1981; Adams and Harris, 1987; Adams and Wake, 1984; Adams, Comyn and Wake, 1997). The research results show that presence of the adhesive spew fillet can reduce the peak peel and shear stresses in the adhesive layer and therefor can improve the joint strength. Reduction in the peak stresses may also be dependent of the shape and size of the spew fillet. To illustrate the effect of spew fillet on the peak stresses in adhesive bonded joints, let us consider the same double lap joint as Figure 2.23a with a triangular spew fillet as shown in Figure 2.26a. Three meshes are generated using 4-node quadrilateral and 3-node triangular elements for the spew fillet area corresponding to the three meshes described in the previous subsection.

Figures 2.27-29 show the peel, shear and longitudinal stress distributions near the overlap end along the centerline of the adhesive layer for the double lap joints with and without adhesive spew fillet. Clearly the peak peel and shear stresses in the joint with spew fillet are smaller than those without spew fillet. Unlike the peel and shear stresses, an increase in the peak longitudinal stress is noted when adhesive spew fillet is present. This means that more load is transferred near the overlap end.

Figures 2.30-32 plot the longitudinal, peel and shear stresses at the adhesive interface corner (see Figure 2.23a) versus the element sizes to demonstrate the stress convergence for joints with and without spew fillet. It is clearly shown that when the element size is decreased from 0.2 to 0.05 mm the stresses at the adhesive interface corner increase when there is no spew fillet. However, when the spew fillet is present, these three stresses become constant as the size of the element is reduced. This example also illustrates the effect of the spew fillet in terms of eliminating mesh dependency of the stresses occurred at the adhesive interface corner.

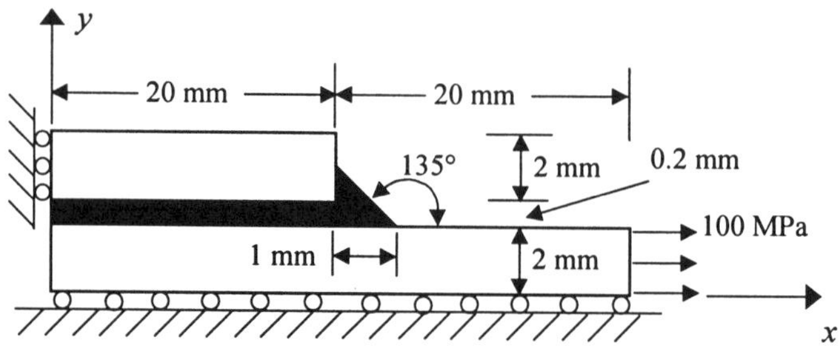

Figure 2.26a Double lap joint with an adhesive spew fillet in tension

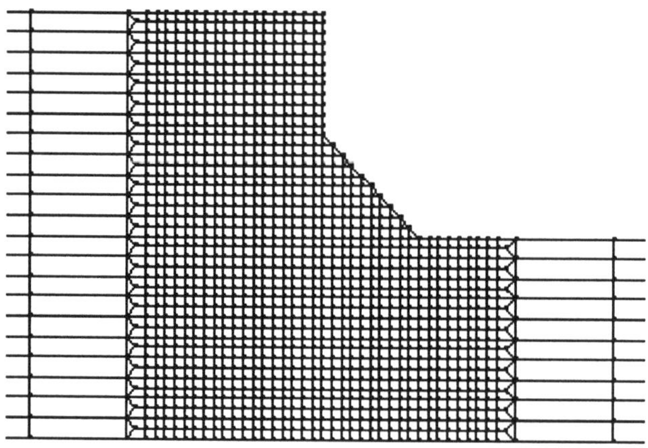

Figure 2.26b Mesh details near the overlap end in the double lap joint with an adhesive spew fillet

It is worth pointing that mesh dependency occurs not only in linear analysis but also in nonlinear analysis. Even when the elastic-plastic adhesive behavior is taken into account in the nonlinear analysis, mesh dependency still exists because the strains, instead of stresses, at the adhesive interface corner are mesh dependent (Sheppard et al, 1998). However, when the spew fillet is included in the finite element analysis model, the mesh dependency of strains can be eliminated similar to the case of linear elastic analysis.

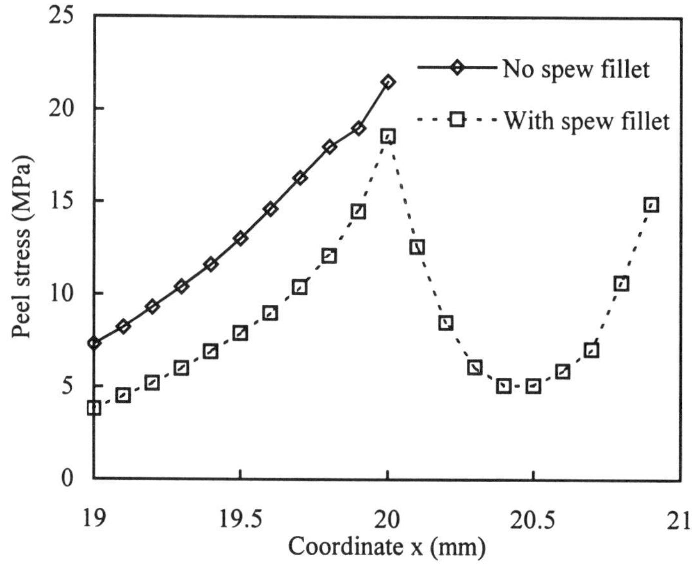

Figure 2.27 Peel stress distribution along the center line of the adhesive layer for the double lap joint with and without spew fillet in tension

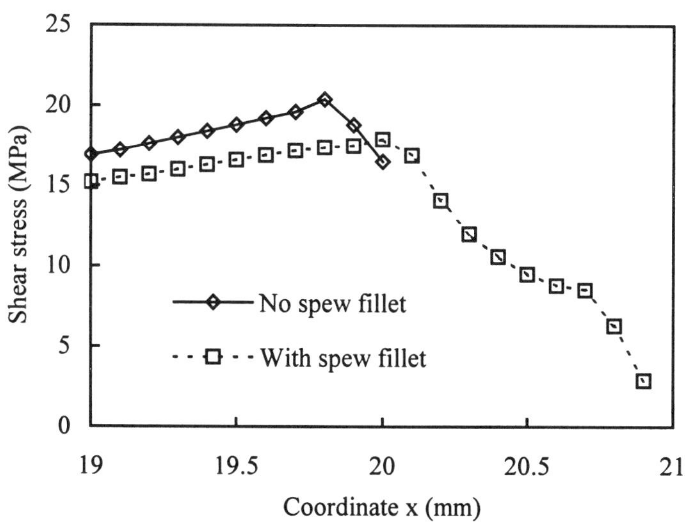

Figure 2.28 Shear stress distribution along the center line of the adhesive layer for the double lap joint with and without spew fillet in tension

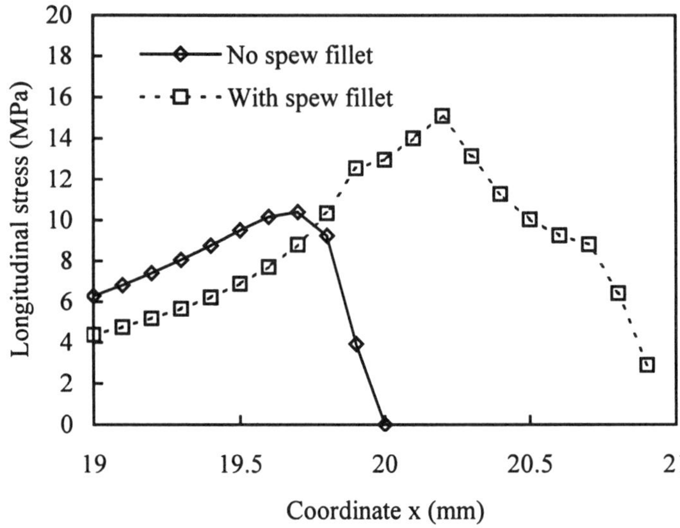

Figure 2.29 Longitudinal stress distribution along the center line of the adhesive layer for the double lap joint with and without spew fillet in tension

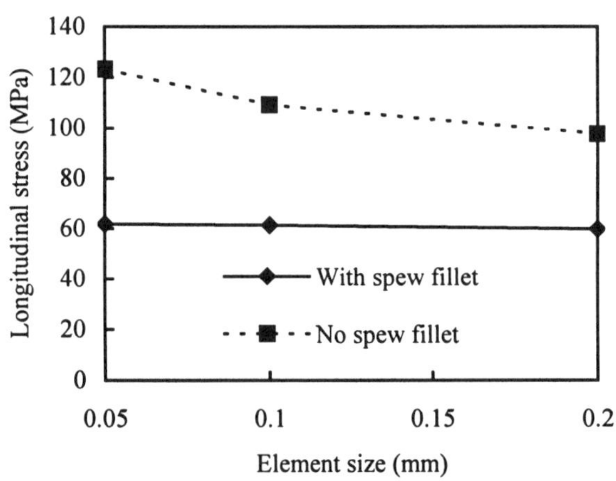

Figure 2.30 Longitudinal stress at the adhesive interface corner of the double lap joint with and without spew fillet in tension versus the element size

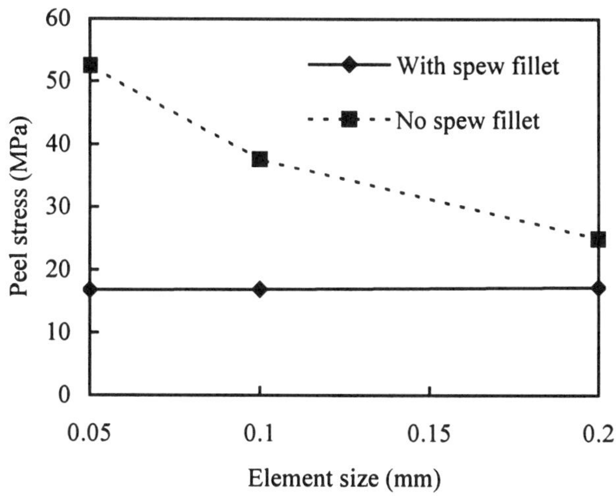

Figure 2.31 Peel stress at the adhesive interface corner of the double lap joint with and without spew fillet in tension versus the element size

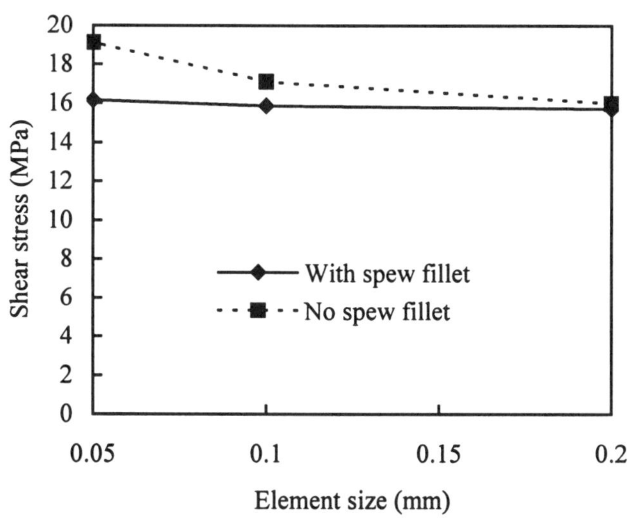

Figure 2.32 Shear stress at the adhesive interface corner of the double lap joint with and without spew fillet in tension versus the element size

(c) Large deformation or rotation and nonlinear material behavior
Some bonded joints, such as double lap, butt and thick adherend joints undergo small deformations and can then be analysed with a reasonably good accuracy using a small deformation formulation. Others like single lap joint, lap shear joint and peel test joint, experience intrinsic large deformation or large rotation even subject to a small load. There exist a large number of finite element analysis results in the literature, which demonstrate the importance of taking into account the large deformation and the nonlinear material characteristics of joining materials, e.g., Crocombe and Adams (1982), Harris and Adams (1984), Picket and Hollaway (1985), Edlund and Klarbring (1993), Edlund (1994) and Tong, Sheppard and Kelly (1995,1998).

To illustrate the effect of large deformation and nonlinear adhesive characteristics on the peak stresses in the adhesive layer, let us consider a single lap joint in tension. Each adherend has a length of 160 mm and a thickness of 2 mm. The overlap length is 60 mm and thus the free length of each adherend is 100 mm. The adhesive thickness is 0.2 mm. The single lap joint is clamped at both ends and is subjected to a tensile stress of 300 MPa across the thickness. The Young's modulus and the Poisson ratio are 70 GPa and 0.3 for the adherend and 2.4 GPa and 0.32 for the adhesive for the linear analysis, and the nonlinear analysis only taking into account the deformation. To include the effect of nonlinear material characteristic, the tensile stress-strain curve depicted in Figure 2.33 is used for the nonlinear analysis case considering both large deformation and nonlinear material characteristics. 2-D Plane-strain finite element analyses are conducted using the Strand 6 (G+D, 1995) by neglecting and taking into account the large deformation and the nonlinear adhesive characteristic. The nonlinear solution procedure applies the applied stress to the joint in 20 increments. Detailed mesh near an overlap end is shown in Figure 2.34.

Figure 2.35 depicts the peak peel and shear stresses at the center of the adhesive element as indicated in Figure 2.34 for the single lap joint for the three types of analyses. These are (a) the linear analysis, (b) the nonlinear analysis including larger deformation only (nonlinear 1) and (c) the nonlinear analysis including both large deformation and material nonlinear characteristics. It is clearly shown that there exists a significant difference between the peak peel and shear stresses in the center of the adhesive element calculated using linear and nonlinear analyses. Such difference becomes noticeable even at a stress level as low as of 30 MPa and becomes significant at high stress level primarily due to large rotation produced in the overlap. This result demonstrates that nonlinear analysis procedure must be used for single lap joints, even subject to a low level of load, in order to predict accurately the stresses in the joint.

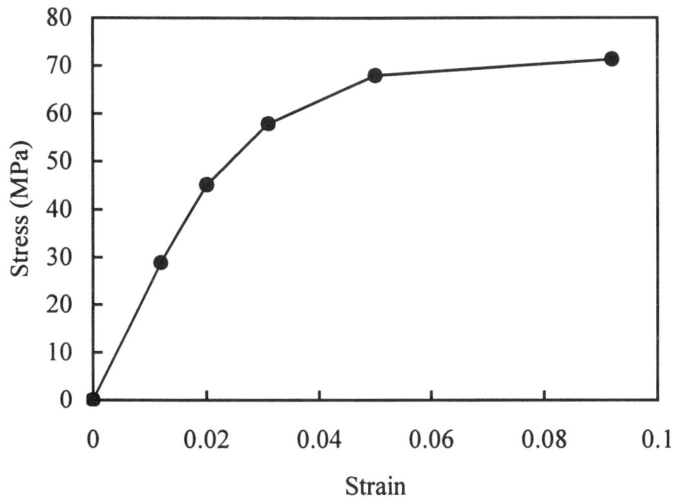

Figure 2.33 Tensile stress-strain curve used in the nonlinear analysis

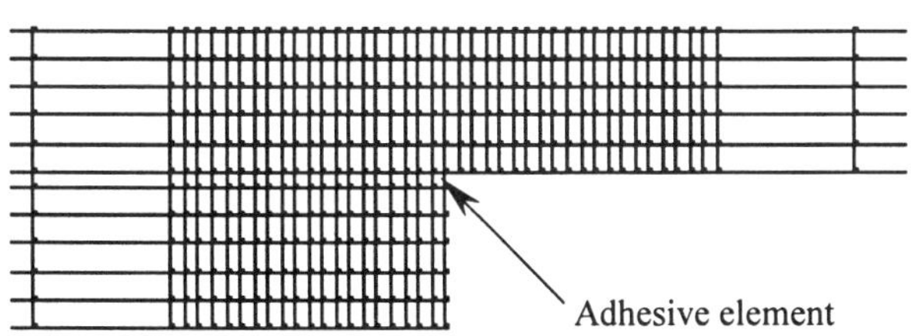

Figure 2.34 Double lap joint with an adhesive spew fillet in tension

Figure 2.35 also shows that there is a remarkable difference between the nonlinear analyses with and without nonlinear adhesive characteristic included. The effect of nonlinear adhesive characteristic on the peak shear stress becomes significant as the applied stress becomes larger than 90 MPa. The effect of nonlinear adhesive characteristic on the peak peel stress is however only noticeable and is not as significant as that on the shear stress.

2.5 CORRELATION OF ANALYTICAL, FEA AND TEST RESULTS

Various analytical and finite element analysis solutions have been developed by many researchers to provide an understanding of the stress distribution in adhesive joints. To gain confidence at the simplified analytical solutions, one thing is to validate analytical solutions by comparing to finite element analysis results.

Both analytical and numerical results show that the peel and shear stresses in the adhesive are not uniform along the overlap length, and their gradients are very steep close to the ends. To confirm this prediction experimentally, a number of testing techniques have been used by many researchers. For example, Renton and Vinson (1977) and Sawyer and Cooper (1981) use the photoelastic technique to measure the stresses in the adhesive layer and adherends for single lap joints with and without preforming. Dillard et al (1989) and Anderson et al (1994) developed a piezoelectric sensor and embedded it in the adhesive layer to measure the average peel stress over the sensor area for a single lap joint and a butt joint. Although the technique requires a dynamic loading and has temperature restrictions, it provided a reasonable point-wise correlation between the measured and the calculated peel stresses. Tuttle et al (1984) embedded strain gauges in the adhesive layer to measure the strain in the bonded joints.

Tsai and Morton (1993) used the Moire fringe technique to measure the displacement fields from a side-view of a single lap joint. A correlation was obtained for the shear stress only in the adhesive between the experimental and the theoretical predictions. Measurement for the peel stress was not in a good agreement with the prediction near the overlap end. Vallat et al (1986) applied the coherent optics to study adhesive joints. Tong, Sheppard and Kelly (1995, 1996) and Tong and Zhuang (1997) proposed an approach that can be used to confirm the peel stress prediction for a bonded double lap joint. In their approach, the surface normal displacement of the outer adherend was measured using the holographic interferometry technique and then used to obtain the peel stress in the adhesive layer.

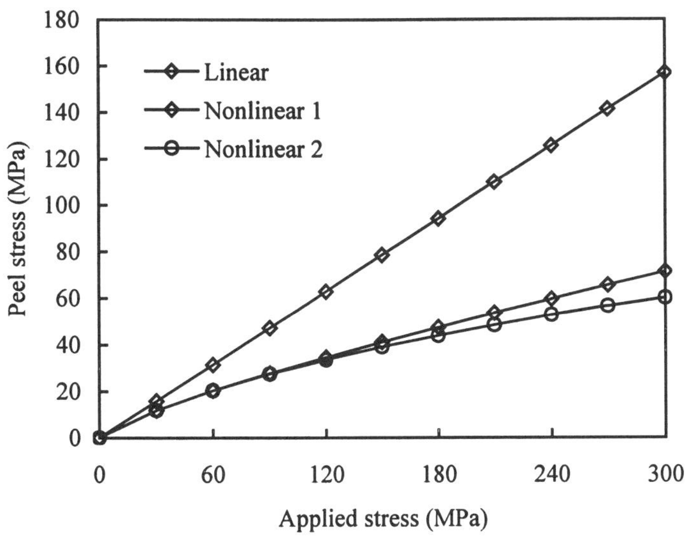

(a)

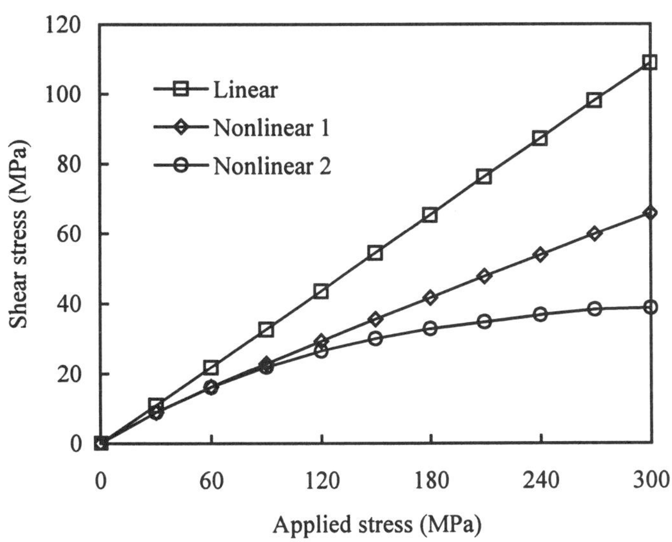

(b)

Figure 2.35 Comparison of peak peel and shear stresses calculated by a linear analysis and a nonlinear analysis taking into account the large deformation and/or nonlinear material characteristic

In this section, we will discuss the comparison of the 1-D analytical solutions and the finite element analysis solutions, and some latest attempt to validate the peel stress distribution in bonded joints. The first example is for adhesive bonded composite double lap joints, and the second example demonstrates the experimental validation of the peel stress distribution in an adhesive bonded doubler.

2.5.1 Peak stresses in double lap joint with composite adherends

Consider an adhesive bonded composite double lap joint subject to axial tension. All adherends are made from the T650/F584 graphite/epoxy unidirectional tape. The inner adherend has a lay-up of [0/90/0/90]s and a nominal thickness of 1.12 mm, and the outer adherend has a lay-up of [0,90]s and a nominal thickness of 0.56 mm. The overlap length is 33.5 mm. The nonlinear stress-strain characteristics of the film adhesive FM300-K (see Figure 2.1b) are used and the nominal adhesive thickness is taken to be 0.15 mm. In the linear analysis case, the Young's modulus and Shear modulus are obtained from the two curves in Figure 2.1b. The mechanical properties of each individual T650/F584 graphite/epoxy ply (nominal thickness 0.14 mm) are tabulated in Table 2.4. Two types of analysis solutions are obtained for this double lap joint. One solution is obtained using the 1-D analysis procedure described in Section 2.3.3, which provides solutions for the adhesive shear and peel stresses, and the axial stress in the central adherend. The other one is obtained using the finite element method. In the finite element analysis, a nonlinear 2-D plane-strain finite element model is used to analyse the stresses and strains in the joint using the MSC/NASTRAN (MacNeal-Schwendler, 1992). Large deformations are considered using the available geometrical nonlinear analysis techniques. Composite adherends are assumed to be linearly elastic and modeled using a ply-by-ply meshing scheme in which a fine mesh is used at the end of adhesive layer as schematically shown in Figure 2.36. The tensile stress-strain curve for FM300-K adhesive as depicted in Figure 2.1b is used in conjunction with the Von Mises yielding criterion to model the nonlinear behaviour of the adhesive layer.

To compare the results calculated using the finite element method and the one dimensional analysis procedure, Figures 2.37-2.39 plot the peak peel and shear stresses in the adhesive element and the longitudinal stress in the surface ply element (see Figure 2.36) computed using both methods. Figure 2.37 shows that the peel or normal stresses in the center of the adhesive and surface ply element are slightly larger than the peel stress computed using the 1-D solution procedure when the applied loads are less than 1050 kN/mm. Figure 2.38 reveals that the shear stress in the center of the adhesive element is smaller than that predicted by the 1-D analysis procedure. The shear stress in the center of the surface ply element is much lower than that those in the adhesive layer and predicted by the 1-D analysis. Figure 2.39 shows that there is a good agreement between the longitudinal stress predicted using the 1-D analysis procedure and that in the centre of the surface ply element. However, a relatively large difference is noted between the 1-D and finite element analysis results when the adherends become thicker (Tong, 1997). This is because the 1-D analysis is based on classical beam/plate theory and thus can give a good stress prediction for joint with thin adherends only.

Table 2.4 Mechanical properties of a T650/F584 graphite/epoxy tape

Longitudinal modulus E_L	156.5GPa
Transverse Modulus E_T and E_Z	15.65 GPa
Poisson ratio ν_{LT} and ν_{LZ}	0.324
Poisson ratio ν_{Tz} (neat resin)	0.35
Shear modulus G_{LT} and G_{LZ}	5.19 GPa
Shear modulus G_{TZ} (neat resin)	1.53 GPa

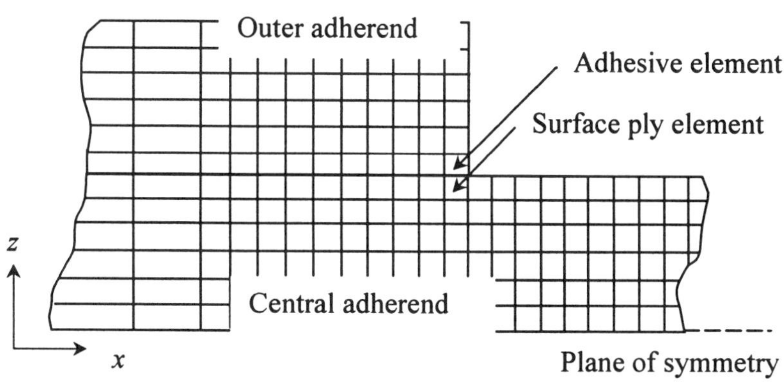

Figure 2.36 Meshing scheme in the vicinity of the adhesive end for a double lap joint

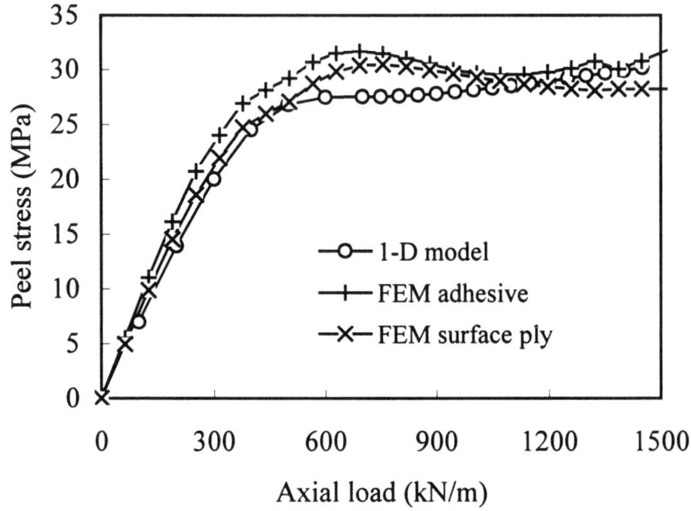

Figure 2.37 Maximum peel stresses (σ_z) vs the applied axial load for the joint

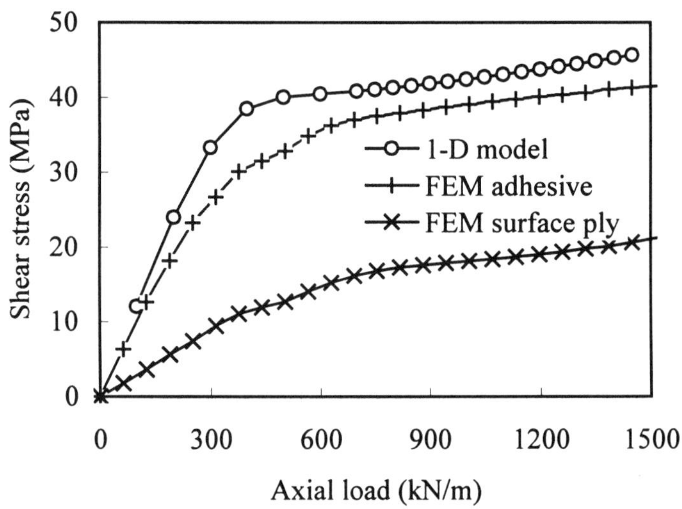

Figure 2.38 Maximum shear stresses (σ_{xz}) vs the applied axial load for the joint

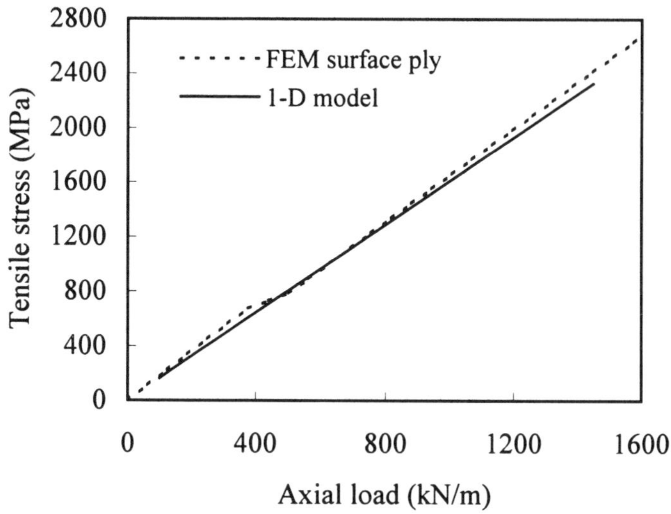

Figure 2.39 Longitudinal stresses σ_x in the surface ply vs the applied axial load for the joint

2.5.2 Peel stress distribution in bonded metal doublers

This section describes some recent results on validation of the peel stress distribution in adhesive bonded metal doubler via measuring the surface normal displacement using the holographic interferomentry technique (Tong and Zhuang, 1997). The peel stress distribution in the adhesive bonded metal doubler is similar to that in bonded double and single lap joints (Tong, Sheppard and Kelly, 1995). An ideal configuration of an adhesive bonded metal doubler is depicted in Figure 2.40a, in which both outer adherends are aligned perfectly to each other. However, it is found that it is extremely difficult to achieve such an exact alignment of the two adherends in adhesive bonding process, and an end mismatch inevitably exists although both adherends are cut to the same length with a specific tolerance. Such doubler with an end mismatch is schematically shown in Figure 2.40b.

An adhesive bonded metal doubler specimen is prepared. The specimen is made of Aluminium 2024-T3, and the adhesive used is FM300-K film adhesive. The material properties, including the nonzero values of the stress-strain curves, of Aluminium 2024-T3 and film adhesive FM300-K are tabulated in Table 2.5. A tensile load P of 7 KN is applied. The dimensions of the specimen with an end mismatch are measured as: outer adherend thickness t is 1 mm, inner adherend thickness $2t$ is 2 mm, adhesive thickness t_a is 0.19 mm, overlap length $2l$ is 60 mm, width of the specimen is 25 mm.

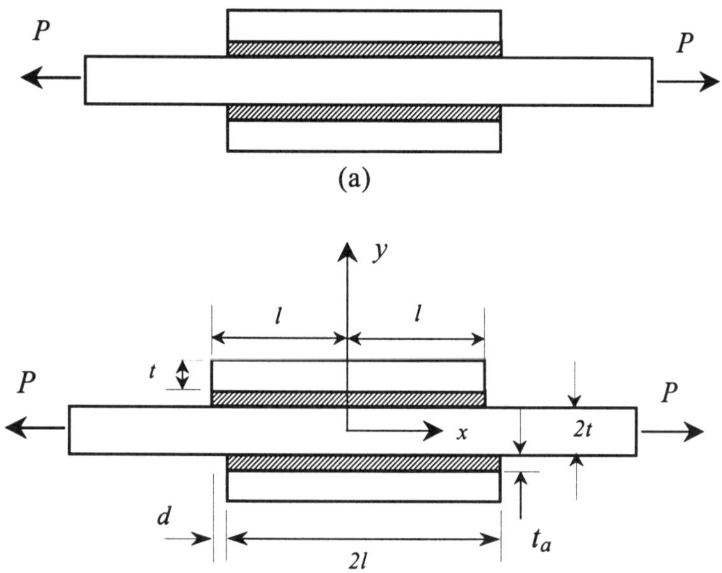

(b)

Figure 2.40 Schematic of adhesive bonded metal doublers (a) an ideal doubler (b) a doubler with end mismatch (Tong, Sheppard and Kelly, 1996).

Finite element analyses are performed using the MSC/NASTRAN (MacNeal-Schwendler, 1992). 2-D Plane-strain models are developed for the doubler specimen using the four-node isoparametric quadrilateral elements. The whole joint specimen is modeled because presence of the end mismatch nullifies the geometrical symmetry. One element is used through the adhesive thickness of the model, ie. y-direction. Following the meshing strategy described in the previous section, a relatively coarse mesh is used along the length (i.e. x-direction, see Figure 2.40(b)) and a fine mesh is employed in regions where the stress gradients are expected to be high. The elastic-plastic material nonlinear analysis is incorporated by assuming that both the adhesive and the adherend are isotropic and using the data given in Table 2.5. The Von Mises yield criterion is used for both the adhesive and adherends.

Table 2.5 Material properties of film adhesive FM300-K and aluminum adherend 2024-T3

Materials	Adhesive (FM300-K)	Adherend (2024-T3)
Young's modulus	2.515 GPa	70.34 GPa
Poisson's ratio	0.36	0.3
σp_1 MPa	30.2	338
εp_1	0.012	0.0048
σp_2 MPa	45.1	355
εp_2	0.0201	0.006
σp_3 MPa	57.8	362
εp_3	0.031	0.0098
σp_4 MPa	67.9	-
εp_4	0.05	-
σp_5 MPa	71.3	-
εp_5	0.092	-

By varying the end mismatch *d* (see Figure 2.40b) from 0 to 1 mm, its effect on the peel stress and shear stress as well as the surface normal displacement are addressed. Figure 2.41a shows the peel stress distribution along the adhesive layer in the overlap for the joints with an end mismatch of 0 and 1 mm. Obviously, when there is no end mismatch the peel stress distribution is symmetric about the midpoint, and when there is an end mismatch of 1 mm, the peel stress at the left end decreases while the peel stress at the other end increases. The maximum increase and decrease are depicted in Figure 2.41b for a variety of doublers with the end mismatch ranging from 0 to 1 mm. The maximum increase and decrease are about 15% and 24% of the peak peel stress relative to the perfectly aligned specimen.

Figure 2.42a shows the shear stress distribution in the adhesive layer along the overlap for the joints with the end mismatch being 0 and 1 mm.

The variations of the peak shear stresses at both ends of the joints with the end mismatch ranging from 0 to 1 mm are depicted in Figure 2.42b. The maximum increase is about 5%, while the maximum decrease is about 9%. It is evident that the end mismatch has less influence on the shear stress than it has on the peel stress when the end mismatch varies from 0 to 1 mm.

Figure 2.43 shows the surface normal displacement distributions along the overlap for the joints with the end mismatch being 0, 0.5 and 1 mm. For the specimen tested, an end mismatch of 0.1 mm was found. It is shown that a small end-mismatch between the top and bottom doublers due to the manufacturing tolerance has an important effect on the distribution pattern of the out-of-plane surface displacement. An end-mismatch produces a local bending moment that can be approximated as a constant across the overlap length. The bending moment creates an additional out-of-plane displacement similar to a single lap joint. Thus the total normal displacement can be approximated as the sum of the displacement of the perfect joint and this additional displacement caused by the constant rotation. For the specimen with a small end mismatch, this extra surface normal displacement can be approximated as a rigid body rotation by an angle about the midpoint. The angle can be determined as the slope of the surface normal displacement at the midpoint along the overlap. Similar to the surface normal displacement caused by the Poisson's ratio effects, this extra surface normal displacement needs to be subtracted from the total displacement.

To calculate the peel stress using the surface normal displacement, Tong, Sheppard and Kelly (1995, 1996) developed the following equation for a perfectly aligned and a mismatched doubler:

$$\sigma_{pl} = \frac{EE_a}{(E + kE_a)t_a}(w_{sf} - w_{ar} - w_{pn}) \tag{2.42}$$

where w_{sf} is the out-of-plane surface normal displacement of an imperfect specimen, w_{ar} is the out-of-plane displacement due to rotation of the overlap, k is a constant between 0 and 2, and w_{pn} is the Poisson's contraction given by (Tong, Sheppard and Kelly, 1995)

$$w_{pn} = \frac{t\sigma}{2Et + E_a t_a}(2\nu t + \nu_a t_a) \tag{2.43}$$

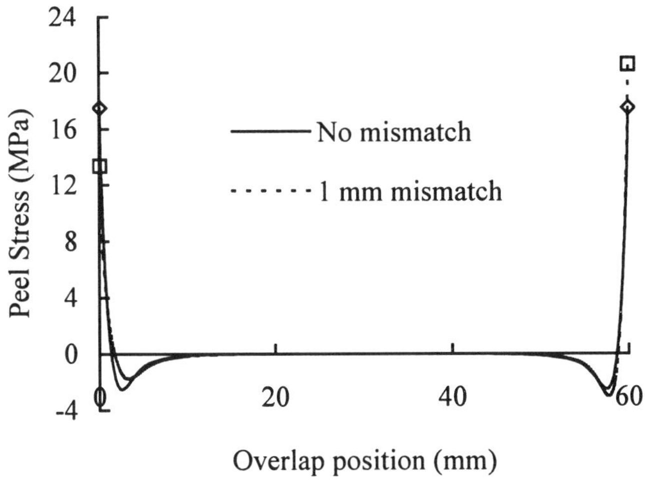

Figure 2.41a Effect of end mismatch on peel stress distribution along the overlap

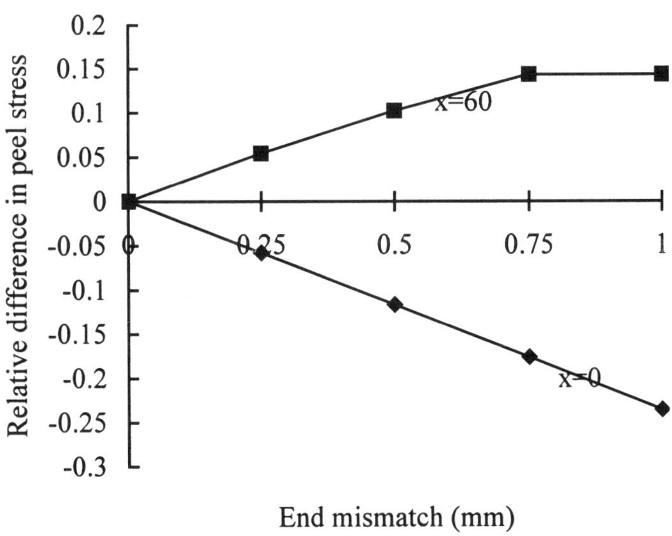

Figure 2.41b Effect of end mismatch on maximum peel stresses at x=0 and x=60 mm

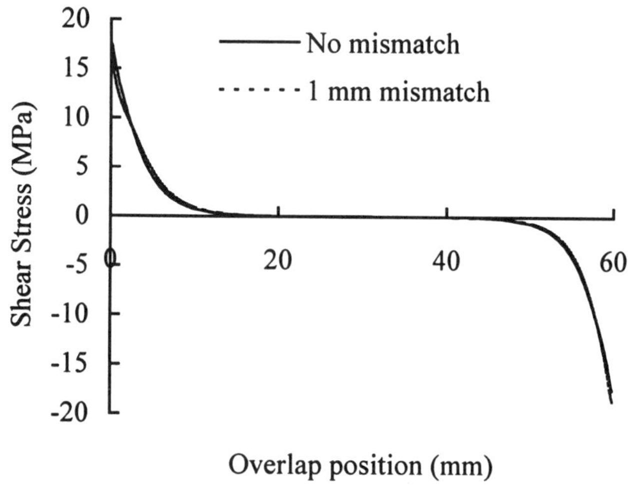

Figure 2.42a Effect of end mismatch on shear stress distribution along the overlap

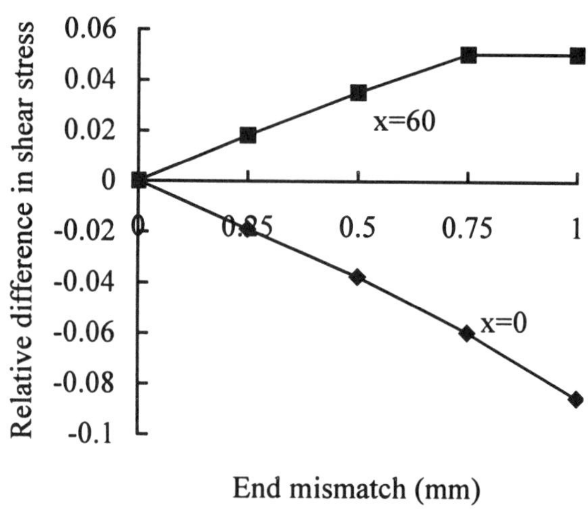

Figure 2.42b Effect of end mismatch on maximum shear stresses at x=0 and x=60 mm

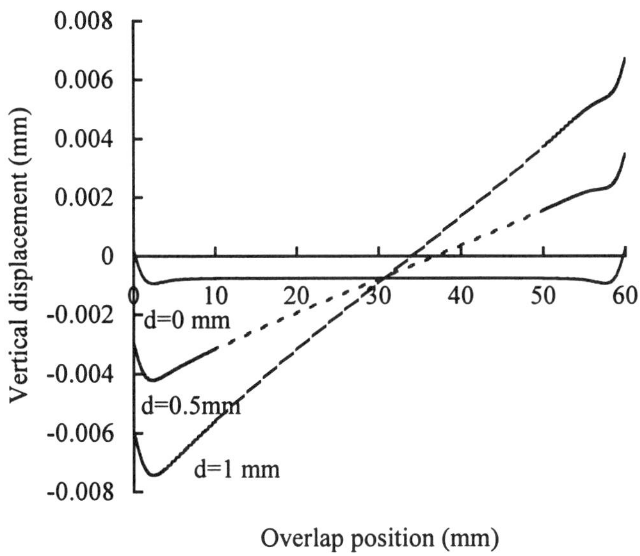

Figure 2.43 Effect of end mismatch on the distribution of the surface normal displacement along the overlap

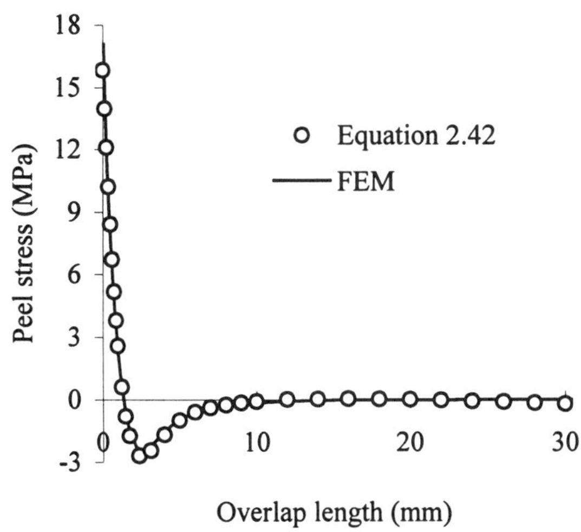

Figure 2.44 Comparison between peel stresses predicted by FEM and equation 2.42 with k=0.

where $\sigma = P/2bt$ and E, ν and E_a, ν_a are the Young's modulus and the Poisson's ratio for the adherend and the adhesive respectively.

To validate equation (2.42) numerically, we can compare the peel stresses directly computed from the finite element analysis with those computed using equation (2.42) and the surface normal displacement of the finite element analysis. Figure 2.44 compares the peel stress distribution given by the finite element analysis and that computed using equation (2.42) with k=0 for the specimen with an end mismatch of 0.1 mm. A reasonable correlation is noted between the stresses given by the finite element analysis and equation (2.42).

To validate the peel stress distribution, the surface normal displacement of the outer adherend of the doubler is measured using a portable holographic interferometry technique system (Tong and Zhuang, 1997). The technique uses single-beam reflective holographic interferometry to obtain interferograms of displacement between two load states. The test set-up contains a laser unit and a holographic plate. The laser unit, mounted on a conventional camera tripod, consists of a spatial filter, a beam expander and a laser source (Zhuang et al, 1994). A holographic plate is attached to one side of the bonded joint at two locations using double-sided adhesive tape. The laser light absorbed by the plate on its first pass represents the reference beam while the light reflected from the surface of the measured object is the object beam. Through double exposure, the change in the out-of-plane surface displacement between two load steps can be picked up by the fringes in the hologram. Thus the relative out-of-plane surface movement can be computed using the following equations:

For bright fringe: $$W = \frac{n\lambda}{2} \tag{2.44a}$$

For dark fringe: $$W = \frac{(2n+1)\lambda}{4} \tag{2.44b}$$

Where W denotes the out-of-plane surface displacement, n is an integer and λ is the wavelength of the laser light (0.633 μm). However, determination of the absolute displacement requires the absolute displacement of a fringe and the sign of the gradient fringe to be known. In general, it is difficult to define the absolute displacement and to determine the sign of each fringe without a physical understanding of the problem.

However, a closing U-shape contour can indicate the change of the gradient associated with the maximum (hill) or minimum (valley).

To minimise the effect of the end mismatch, a rig to hold the specimen was used to shorten the free length of the central adherend in order to reduce the local bending effect. After careful installation of the specimen on the Instron machine to ensure appropriate alignment, the specimen is loaded and two exposures are taken at different load levels. The first exposure is taken at a load level of 1.0 KN, and the second exposure taken at 7.0 KN. The holographic plate is then developed and dried to obtain the interferometric fringes. Figure 2.45 depicts the holographic fringe image over half of the overlap of the joint specimen. The fringes are then analyzed using equation (2.44) and the surface normal displacement is obtained. In the following results, the displacement distributions are only given for half of the specimen along the overlap centerline and along the overlap edge as shown in Figure 2.45.

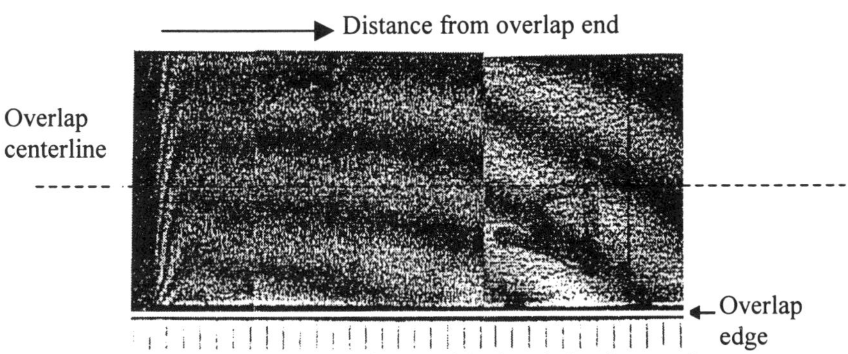

Figure 2.45 Holographic image of a double lap joint in tension (redrawn from Zhuang et al, 1994)

Figure 2.46 depicts the surface normal displacement distribution measured using the holographic technique along the overlap centerline as shown in Figure 2.45 and also that computed using the finite element method. A reasonable correlation between the test data and the numerical prediction is observed. The peel stress is computed using equation (2.42) and the measured surface normal displacement, and then compared with that calculated using the finite element method. These comparisons are shown in Figure 2.47 with k=0, which shows that there exists a good correlation between the stresses measured and those computed using the finite element analysis method.

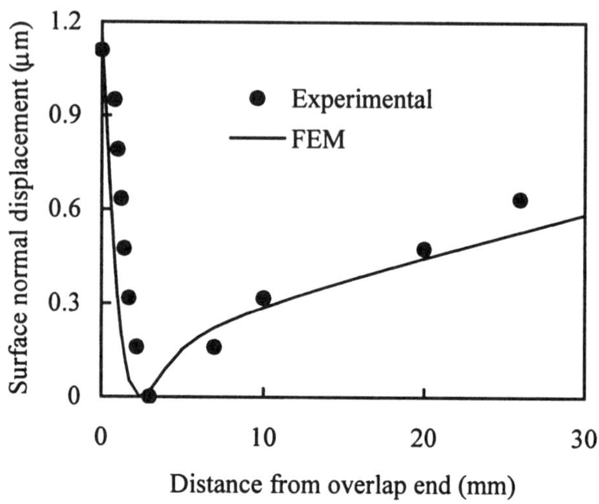

Figure 2.46 Experimental and numerical correlation for Surface displacement along the overlap centerline (see Figure 2.45)

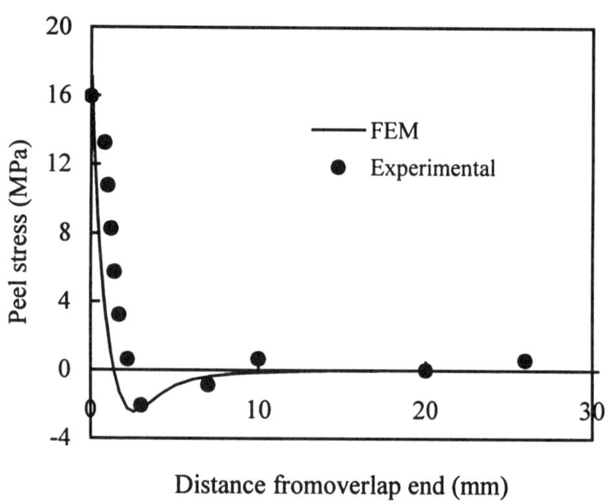

Figure 2.47 Experimental and numerical correlation for peel stress (k=0) along the overlap centerline (see Figure 2.45)

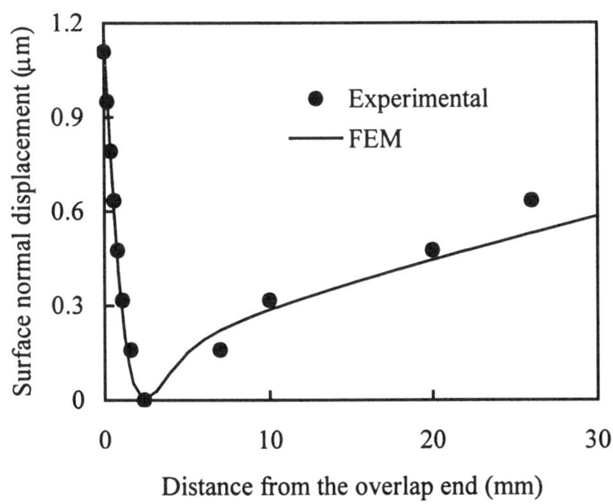

Figure 2.48 Experimental and numerical correlation for the normal surface displacement along the overlap edge (see Figure 2.45)

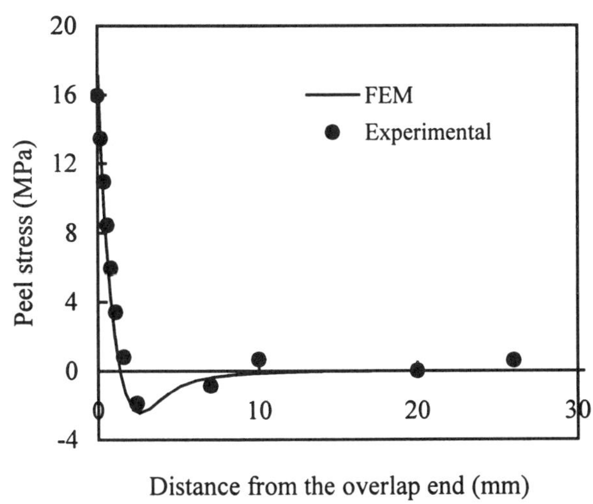

Figure 2.49 Experimental and numerical correlation for peel stress (k=0) along the overlap edge (see Figure 2.45)

Figure 2.48 depicts the surface normal displacement distribution along the overlap edge as shown in Figure 2.45 measured using the holographic technique and that computed using the finite element method. A better correlation between the test data and the numerical prediction is observed compared to that along the overlap centreline. The peel stress is computed using equation (2.42) and the measured surface normal displacement, and then compared with that calculated by the finite element method. These comparisons are shown in Figure 2.49 with k=0. Once again, a good correlation is noted for the peel stress.

3

Failure criteria and strength prediction

3.1 INTRODUCTION

There are in general two approaches in structural design and material selection. One is a strength of materials based approach, and the other is a fracture mechanics based approach. In the first one the applied or anticipated design stress or strain is compared to the strength or the maximum strain of the candidate material, and then the material is assumed to be adequate if its strength is greater than the expected applied stress. In the second one, there are three important variables, the applied stress, the flaw size and the fracture toughness, and the critical combinations of these three variables are quantified using fracture mechanics to determine the strength and service life of a structure. A structural bonded joint is often one of the most important part in an engineering structure. Proficient design of structural joints is essential for load-carrying structures because joints often impose a significant weight penalty and often establish structural efficiency. To design efficient joints and to best utilize the advantages of adhesive bonding, it is imperative to develop simple and reliable methodologies that can provide accurate and reliable estimations on joint strength and service life. In this chapter, we will address this issue from both aspects: the strength of materials and the fracture mechanics failure criteria and examine their application to predicting joint strength and service life.

Structural bonded joints can fail at different location and in a variety of failure modes. Failure can occur or initiate in the adhesive or in the adherend, depending on the geometrical configuration, the materials of the adherends, the adhesive as well as the manufacturing procedure. It is difficult to definitely describe and define all the possible failure modes of adhesive bonded joints. This is particularly true for adhesive bonded composite joints. There is an ASTM standard D5573-94 describing the standard practice and method for classifying, identifying, and characterizing the failure modes in adhesive bonded fiber-reinforced-plastic joints. To present various existing failure criteria, we classify the failure modes of adhesive bonded joints into the following four general groups: (i) adherend failure due to bending, tension or compression etc, (ii) adherend-adhesive interfacial failure, (iii) cohesive failure, and (iv) out-of-plane adherend failure due to delamination in composite adherend. Although these four groups of failure modes have been identified, they can take place in an adhesive bonded joint simultaneously in a mixed or combined form.

Figure 3.1 depicts schematically some possible cohesive and adhesive failure modes in an adhesive bonded single lap joint with and without adhesive spew fillet. Figure 3.1(c) and (d) indicate the possible loci where adhesive failure can initiate and the possible crack growth directions (see Adams, Comyn and Wake, 1997). The crack formation procedure in ductile adhesive can be seen from Figure 3.2, which shows void nucleation, growth, and coalescence in ductile materials. Figure 3.1(e)-(g) depicts schematically the possible crack growth paths in the overlap lengthwise direction, the first one is within the adhesive between the two adherend-adhesive interfaces, the second is along the adherend-adhesive interface, and the third is a wavy crack path. For example, Chai (1987) reported a wavy crack path in an adhesive layer between aluminum adherends. The crack jumps periodically from one interface to the other across the adhesive layer.

Adherend-adhesive interface failure often takes place at a macro scale when surface preparation or material quality is poor. In this book it is assumed that all joints are manufactured to specifications, and consequently the adherend-adhesive interface failure can not be predicted using any theoretical techniques.

When joining organic composite members, particularly thin-walled composite members, an adhesive bond is often selected as the preferred way of structural joining because adhesive itself is basically the same polymer as resin matrix. However, due to the laminated nature of composite members and the relative weakness in the through-the-thickness direction, the failure mechanism of adhesive bonded composite joints becomes more complex than that of a metal-to-metal joint. Figure 3.3 shows two typical failure modes, net section fracture and delamination, for an adhesive bonded composite single lap joint (Hoskin and Baker, 1986; Kairouz and Matthews, 1993, Tong, Sheppard, Kelly and Chalkley, 1998).

The delamination failure is also schematically shown in Figure 3.4 with two possibilities: the surface ply fracture (Figure 3.4(c)) and further growth of delamination (Figure 3.4(d)). Apparently, more factors need to be considered in strength prediction of composite joints due to complexity in material properties.

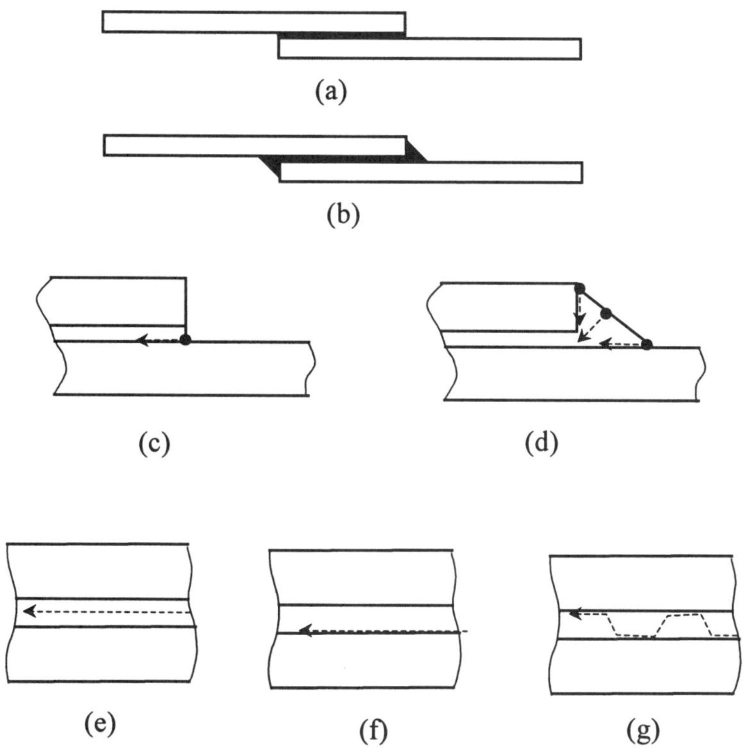

Figure 3.1 Some typical adhesive failure modes of adhesively bonded metal single lap joints: (a) a single lap joint with no adhesive spew fillet, (b) a single lap joint with triangular spew fillet, (c) possible failure location in single lap joint with no spew fillet, (d) possible failure loci and directions in a single lap joint with spew fillet, (e) crack propagating in adhesive layer, (f) crack propagating along or near the adhesive and adherend interface, and (g) wavy crack growth along the two interfaces.

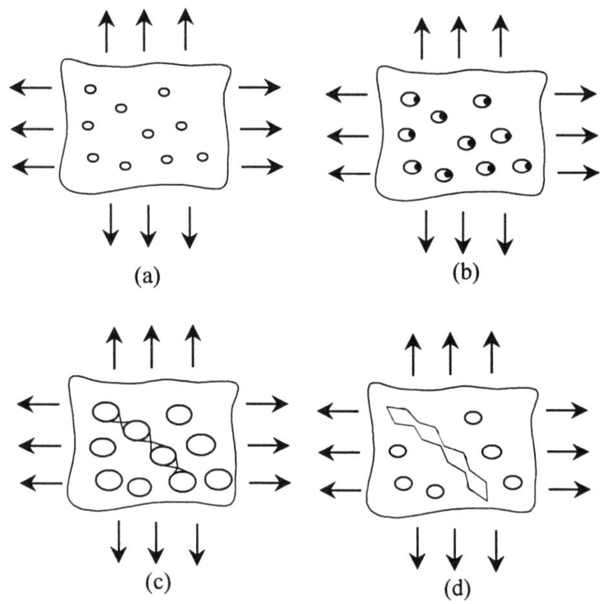

Figure 3.2 Void nucleation, growth, and coalescence in ductile materials

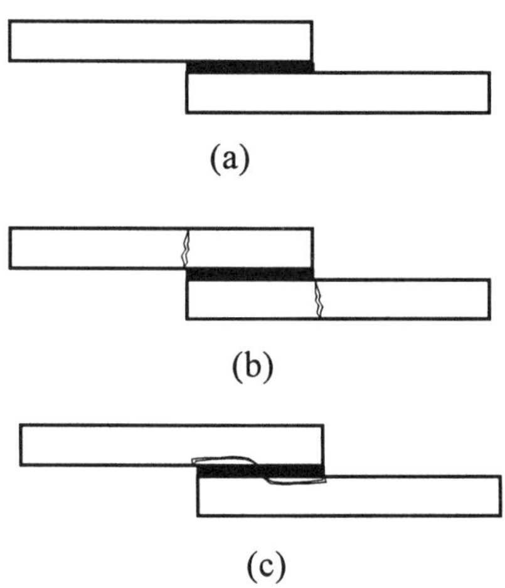

Figure 3.3 Typical failure modes in an adhesive bonded composite single lap joint

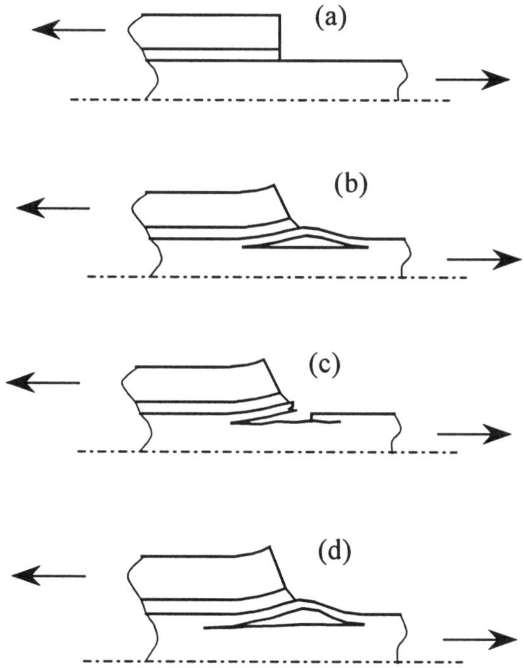

Figure 3.4 Peel stress induced failure in composite lap joints

Accurate prediction of bonded joint strength and service life is one of the most challenging tasks in their structural design. First of all, it is not an easy job to determine with certainty the accurate geometry of the irregular adhesive spew fillet as it can significantly change the stress concentrations at the adhesive ends. Secondly, it is difficult and expensive to characterize the adhesive properties, such as stress-strain curves and material allowables including fracture toughness, especially when environmental effects are present. One reason for this is the existence of discrepancy of a property measured using various test methods although there are a number of standard testing methods as described in Chapter 1. One example is the shear stress-strain curves measured using the napkin ring test method and two different thick adherend test set-ups as given in Figure 2.3. Lastly, but not least, there is a lack of complete understanding and accurate modeling methodologies of failure mechanisms in various bonded joints. Joint failure mechanism includes onset of initial failure, such as yielding, fracturing, and cracking, followed by stable growth of crack(s) or development of yielding zone(s), and ultimately fast or catastrophic crack propagation. The failure mechanism becomes even more complex when one of the joining member is laminated composite material. In this case, additional affecting factors, such as lay-up sequence, fiber orientation, material properties of each ply, are involved in joining composite members. When damage exists in an adhesive bonded joint, particularly in a composite joint, prediction of the effect of damage on the joint strength and service life becomes another important and challenging issue related to analysis and design of structural bonded joints, which will be discussed in Chapter 4.

There are a large variety of failure criteria in the literature that have been used in the strength prediction of adhesive bonded joints. As mentioned before, all methods of predicting joint strength and service life can be grouped into two major areas, one is the strength of materials based approach and the other is the fracture mechanics based approach. The former one has often been used to predict static or dynamic joint strength, while the latter one has been used frequently to predict static strength and service life for bonded joints.

This chapter aims to present some of the failure criteria available in the literature from both areas and to discuss their applications to selected joints. It is certainly not possible to include all the existing failure criteria due to length limitation. However, keen readers are recommended to read the articles on failure criteria of joints and composite materials, for example, Matthews et al (1982), Adams (1989), Beaumont and Schultz (1990) and Sheppard, Kelly and Tong (1998), and to read articles on fracture mechanics based failure criteria, for example, by Hutchingson and Suo (1992).

3.2 STRENGTH OF MATERIALS CRITERIA

The strength of materials based criteria refer to those criteria comparing stresses and/or strains with the corresponding material allowables, such as material strengths. These criteria have been used to predict static strengths of a variety of joints or conditions at which structural integrity fails. Adhesive bonded structural joints can fail in different failure modes and at different locations depending on many factors, such as material combinations, loading environment etc. However, for quality bonded structures, joint failure occurs either in adhesive or in adherends (or substrates) or both. In this section, failure criteria based on strength of materials are classified into two groups, i.e. adhesive failure criteria and adherend (or substrate) failure criteria.

3.2.1 Adhesive failure criteria

Failure criteria developed for isotropic materials have been utilized to predict adhesive failure in adhesive bonded structures. There exist a number of new failure criteria that are developed specifically for predicting adhesive failure. In general, all failure criteria can be grouped into two categories: point-based and zone-based failure criteria. In the point-based criteria, the failure conditions are satisfied at an individual critical point; while in the zone-based failure criteria, the failure conditions are satisfied over a zone of fixed-size. For an adhesive bonded metal-to-metal lap joint adhesive failure is typical. This is because metal adherends are often stiffer and stronger than adhesives, and therefore large plastic deformation takes place in the adhesive.

(a) Maximum principal stress or strain criteria:

In this failure criterion, adhesive failure is assumed to occur when the maximum principal stress or strain attains its limiting value. This failure criterion has been used, in conjunction with a geometric and material nonlinear finite element analysis, to perform failure analysis of adhesive bonded joints by Harris and Adams (1984), Adams, Atkins et al (1986), Adams and Harris (1987), Adams (1989).

In strength prediction of adhesive bonded joints, the finite element method has been widely used in conjunction with a variety of failure criteria (see, e.g. Adams, 1989; Czarnocki and Piekarski, 1986; Hollaway et al, 1990; Lee and Lee, 1992, Mori and Sugibayashi, 1990). Unfortunately, the problem with using FEA and indeed the basic phenomenon which makes bonded joint failure predictions very difficult is the presence of both geometric and material discontinuities at the adhesive-adherend interface ends in the joints (Delale, 1984; Hein and Erdogan, 1971). The presence of such discontinuities may cause a stress or strain concentration or even a stress or strain singularity, which can not be eliminated in a full non-linear elastic-plastic analysis. The accuracy of the finite element solutions for strains near the joint end and the strength therefore depends on the refinement of the finite element mesh.

To overcome the FEA singularity problem, a number of researchers have developed the stress singularity parameter approaches to predict joint failure (Hattori, 1991; Groth, 1988). However, in reality true singular points probably do not exist at the ends of the joint due to the fact that the corners at the ends of the joint will not be perfectly square. In addition, the stresses at the ends of the joint will be relieved due to zones of local damage, which can take the form of voids, local crazing, local cracking etc.

(b) Zone-based criteria:
In zone based-criteria, adhesive failure is assumed to occur when a stress or strain component exceeds its limiting value over a finite zone or area. For example, adhesive failure is said to occur when the maximum tensile or shear stress exceeds its tensile or shear strength over a finite zone. Unlike point-based failure criteria, zone-based criteria introduce a new parameter, the size of the critical zone.

Crocombe (1989) proposes that global yielding of the entire adhesive layer be used as the failure criterion. This criterion avoids many of the inherent problems with joint strength predictions. Unfortunately, this approach is not conservative for many joints because local failure could occur before global yielding.

To overcome these problems, a number of researchers have proposed using a fracture mechanics based approach to predict bonded joint failure. In fracture mechanics, a single sharp macroscopic crack is assumed to exist in the material and fracture occurs due to the propagation of this crack. This part of the work will be discussed in section 3.3.

Recent research results show that failure of polymer structural components and adhesive bonded joints initiate from a zone of damaged material (Knauss and Losi, 1993; Laksimi et al 1994; Chai, 1993a,b). Based on these observations, it seems that, rather than trying to zoom in on the singular points at the joint ends, failure should be predicted by backing away from the singularities and studying the stresses/strains over a finite zone at the end of the joint.

McClintock (1958) suggested that failure occurs in an elastic-plastic, non-work-hardening material in pure shear when a critical shear strain is attained over a critical volume of material. Clark and McGregor (1993) proposed that bonded joint failure could be predicted when the maximum principal stress calculated over a finite zone exceeded the ultimate tensile stress of the adhesive. John (1994) and John et al (1991) proposed that failure occur in bonded composite joints when the adhesive over a critical portion of the joint exceeds the shear yield stress. Sheppard, Kelly and Tong (1998) proposed a damage zone model for the failure analysis of adhesive bonded joints. Their results show that the damage zone model can be used to predict failure strengths and loci of various bonded joints with a reliable engineering accuracy. In addition, the problem of finite element mesh dependence due to the joint singularity is removed as failure is assumed to occur when the material over an area reaches its critical condition.

Also in recent years Continuum Damage Mechanics (CDM) has been developed to describe the progressive degradation of a material prior to the initiation of macrocracks (Simo and Ju, 1987; Edlund and Klarbring, 1993; Edlund, 1994).

(c) Strain energy density based criteria:
The strain energy density is defined as

$$w = \int_0^{\varepsilon_{ij}} \sigma_{ij} d\varepsilon_{ij} \tag{3.1}$$

where σ_{ij} and ε_{ij} are the stress and strain tensors, respectively. Adhesive failure is assumed to occur when the strain energy density at any location in the adhesive exceeds a limiting value. This limiting value can be defined as the area under a tensile or shear stress-strain curve.

Based on an ideal elastic-plastic and bilinear plastic models, Hart-Smith (1973a) proposed a simple formula that expressed the failure load of an adhesive bonded double lap joint in terms of the strain energy density. This simple formula of the bond shear strength provides an efficient tool for designing bonded joints (Thrall (1979); Jones et al. (1993)), and bonded repairs (Hart-Smith (1988); Baker and Jones (1988)), etc. Tong extended the work by Hart-Smith by considering a true stress-strain curve for double lap (1994), single lap and lap shear joints (1996, 1998a).

3.2.2 Adherend failure criteria

Adherends can be made of either metal or composite materials. For metal adherends, various plastic-yielding criteria described in Section 2.4 can be used to predict adherend failure. For composite adherend, adherend failure can occur either in the form of net sectional fracture due to high bending moment at an overlap end (see Figure 3.3(b)), or in the form of interlaminar delamination (see Figure 3.3(c) and 3.4) due to high peel stress and low interlaminar strength. ASTM D5573-94 describes the standard practice and method for classifying, identifying, and characterizing the failure modes in adhesive bonded fiber-reinforced-plastic (FRP) joints. Since the invention of composite materials, failure criteria for composite materials have received extensive study from numerous researchers. Although it is still not fully resolved, there exist a number of reviews on failure criteria of composite materials, for example, Rowlands (1985), Nahas (1986) and Beaumont and Schultz (1990). Various failure criteria have been applied to predict strengths of adhesive bonded composite joints (Matthews et al, 1982; Adams, 1989). In this subsection, we will discuss the failure criteria that may be used to predict the net section fracture and interlaminar delamination.

(a) Failure criteria for net sectional fracture:

For in-plane failure of laminated adherends, there are a variety of failure criteria. For examplc, thc maximum stress criterion, the maximum strain criterion, the Tsai-Hill criterion and the Tsai-Wu tensor criterion (see, Rowlands (1985), Nahas (1986). Beaumont and Schultz, 1990).

Maximum Stress Criterion: failure occurs in a ply when any stress in principal material directions exceeds the respective strength. When the two principal normal stresses are in tension, i.e. positive

$$\sigma_L < X_t, \qquad \sigma_T < Y_t, \qquad |\tau_{LT}| < S \tag{3.2a}$$

and when the two principal normal stresses are in compression, i.e. negative

$$\sigma_L > X_c, \qquad \sigma_T > Y_c \tag{3.2b}$$

Tsai-Hill Criterion: For the case of plane stress, failure is assumed to have occurred in a unidirectional lamina when the following inequality is violated:

$$\left(\frac{\sigma_L}{X}\right)^2 - \frac{\sigma_L \sigma_T}{X^2} + \left(\frac{\sigma_T}{Y}\right)^2 + \left(\frac{\tau_{LT}}{S}\right)^2 < 1 \tag{3.3}$$

where σ_L, σ_T and τ_{LT} are the longitudinal, the transverse and the in-plane shear stresses. *X*, *Y* and *S* are the longitudinal, transverse and in-plane shear strengths. The subscript *t* and *c* indicate in tension and compression.

(b) Interlaminar failure criteria:

Maximum stress criterion: In this criterion, it is assumed that interlaminar failure in a composite adherend occurs when the normal tensile stress at the interface exceeds its ultimate strength, e.g.

$$\frac{\sigma_z}{Z} = 1 \tag{3.4}$$

where *Z* is the interlaminar tensile strength of the composite adherends. σ_z is the normal tensile or peel stress at the interface. The interlaminar tensile strength is normally assumed to be equal to the in-plane transverse tensile strength for a prepreg laminar. It is noted that the interlaminar tensile strength is often much less than the interlaminar shear strength. Instead of using stress and strength values, tensile strain and ultimate tensile strain values can be used as a maximum strain criterion.

Interactive criteria: In interactive criterion, in addition to the interlaminar tensile stress, the interlaminar shear stress and the longitudinal stress are taken into account simultaneously to predict interlaminar delamination.

The two interactive criteria used by Long (1991) to characterize the interlaminar delamination failure of laminate for adhesively bonded ARALL-1 joints are

$$\frac{\sigma_z}{Z} + \frac{\tau_{xz}^2}{R^2} = 1 \tag{3.5a}$$

$$\frac{\sigma_z^2}{Z^2} + \frac{\tau_{xz}^2}{R^2} = 1 \tag{3.5b}$$

where *R* is through-the-thickness shear strength of the laminate adherends, and τ_{Lz} is the shear stress at the interface. Long (1991) found that the observed failures within the prepreg layer can be accurately described using the first interaction equation for adhesively bonded ARALL-1 single and double lap joints. To take into account the interaction of the stresses in the composite adherends, Tong (1997) assessed the following six interactive criteria were assessed by predicting static strength of adhesive bonded double lap joints.

$$\frac{\sigma_x(\sigma_x - \sigma_z)}{X_t^2} + \frac{\sigma_z^2}{Z^2} + \frac{\tau_{xz}^2}{R^2} = 1 \tag{3.6.1}$$

$$\frac{\sigma_x(\sigma_x - \sigma_z)}{X_t^2} + \frac{\sigma_z}{Z} + \frac{\tau_{xz}^2}{R^2} = 1 \tag{3.6.2}$$

$$\frac{\sigma_x^2}{X_t^2} + \frac{\sigma_z^2}{Z^2} + \frac{\tau_{xz}^2}{R^2} = 1 \tag{3.6.3}$$

$$\frac{\sigma_x^2}{X_t^2} + \frac{\sigma_z}{Z} + \frac{\tau_{xz}^2}{R^2} = 1 \tag{3.6.4}$$

$$\frac{\sigma_x^2 - \sigma_x\sigma_z}{X_t X_c} + \frac{\sigma_z^2}{Z^2} + \frac{\tau_{xz}^2}{R^2} = 1 \tag{3.6.5}$$

$$\frac{\sigma_x^2 - \sigma_x\sigma_z}{X_t X_c} + \frac{\sigma_z}{Z} + \frac{\tau_{xz}^2}{R^2} = 1 \tag{3.6.6}$$

where σ_x is the longitudinal stress in the surface ply, X_t and X_c are the ultimate tensile and compressive strength of the surface ply in the laminate. Equation (3.6.1) is the well-known Tsai criterion (Nahas, 1986), criterion in equation (3.6.3) is a degenerated form of Tsai criterion in which the tensile and compressive strengths are assumed to be identical, and criterion in equation (3.6.5) is the Norris criterion (Nahas, 1986). By replacing the quadratic normal stress term in equations (3.6.1,3.6.3 and 3.6.5) with the linear term, Tong (1997) proposes criteria given in equation (3.6.2), (3.6.4) and (3.6.6). It is believed that through-the-thickness normal tensile or peel stress is a predominant factor in causing interlaminar delamination in the surface ply.

3.3 FRACTURE MECHANICS CRITERIA

Adhesive bonded joints of various geometric configurations have complex stress states and stress concentrations at the bondline termination or near the end of bimaterial interface. These stress concentrations may cause debonding initiation and growth, thus may lead to joint failure. Debonding is a self-similar cracking problem as cracking may occur in the adhesive layer (cohesive cracking), along an interface between adhesive and adherends or in the adhesive layer (interfacial cracking) or in wavy cracking path switching between the two adhesive-adhered interfaces. These three typical types of cracking paths have been schematically shown in Figure 3.1.

Because debonding is a self-similar cracking problem, it is natural to describe the crack initiation and propagation phenomena using a fracture mechanics approach (Anderson et al, 1988). Depending on the geometry and loading conditions of a joint, the complex stress states can cause debond initiation in the following three modes (see Figure 1.6):

(a) Opening mode I, due to tensile stress normal to the plane of debond, yields a G_I or J_I strain energy release rate or a K_I stress intensity factor.

(b) Shearing mode II, due to in-plane shear stress in the cracking direction results in a G_{II} or J_{II} strain energy release rate or a K_{II} stress intensity factor.

(c) Tearing mode III, due to anti-plane shear stress perpendicular to the cracking direction, results in a G_{III} or J_{III} strains energy release rate or a K_{III} stress intensity factor.

Determination of the mixed-mode strain energy release rate or stress intensity factor is not trivial for adhesive joints of various geometries, especially when geometric and material non-linearites may exit. If the mixed-mode energy release rate can be calculated with confidence for an adhesive bonded joint or component, it is possible to predict debond initiation using an adequate failure criterion base on measured fracture toughness. If a relationship between the crack growth rate (crack extension per cycle) and the energy release rate induced at the crack tip exists, it is possible to predict the fatigue failure behavior of an adhesive bonded joint or structural component. The procedure of static and fatigue strength prediction is schematically illustrated in Figure 3.5. There are three basic steps in the procedure, and they are

Step 1: Calculate the mixed-mode energy release rate for a given joint configuration under give loading conditions; and further establish the relationship between the crack length and energy release rate for fatigue strength prediction;

Step 2: Measure the critical values of the energy release rates of mode I and mode II, or the mixed mode for the material system used in the joint or structural components. Further more for the case of fatigue the relationship between the crack growth rate and the energy release rate needs to be established;

Step 3: Apply failure criterion for predicting static strength or the load at which debond initiates and/or for determining the number of cycles to failure.

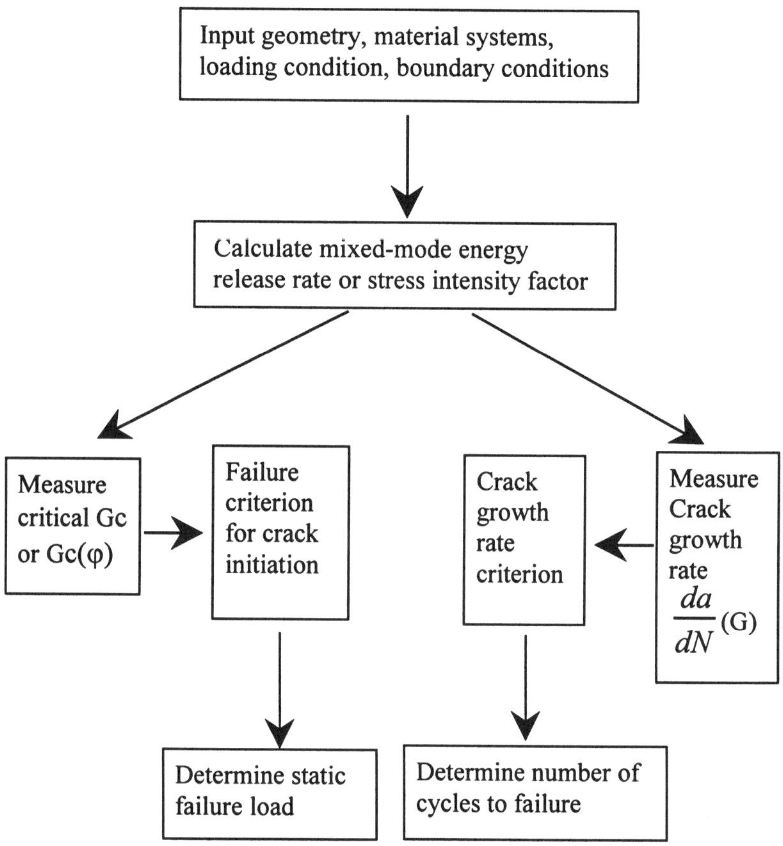

Figure 3.5 Procedure for determining the static strength and fatigue life of adhesive bonded joints

In this section, the fundamentals of the fracture mechanics will be briefly reviewed first to explain the concepts, such as energy release rate, and the methods. Readers are referred to the book by Anderson (1995) for detailed information on fracture mechanics. Illustrative examples will be given to demonstrate various methods in terms of calculation of the energy release rate G or J. Standard Test methods will also be discussed in conjunction with some mixed mode test methods for determining the critical energy release rate and the crack growth rate. Various failure criteria based on fracture mechanics will then be presented.

Adhesives can be classified into two groups according to their stress and strain behavior, one is brittle adhesive and the other is ductile adhesive. As schematically shown in Figure 3.6, stress is proportional to strain for brittle adhesive, and there is little to none yielding or plastic deformation, while in ductile adhesive the stress-strain curve becomes nonlinear, and plastic yielding and deformation can take place. Linear elastic fracture mechanics theory can be used for joints with brittle adhesives, while elasto-plastic fracture mechanics may be used for joints with ductile adhesives.

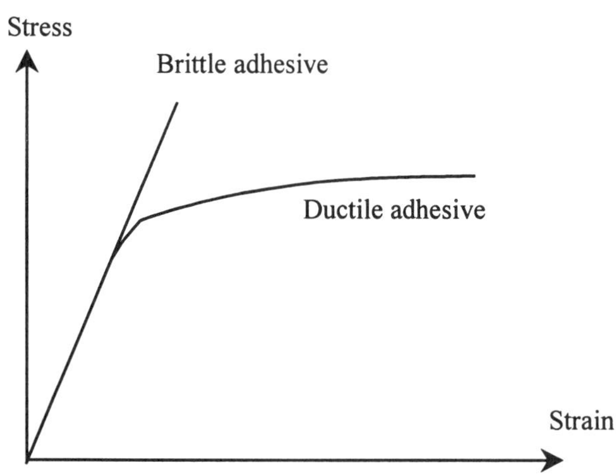

Figure 3.6 Schematic stress-strain curves of brittle and ductile adhesives

For the fracture mechanics approach there are generally two main types of methods, the energy criterion and the stress intensity approach. The energy approach postulates that crack propagation occurs when the energy available for crack growth is sufficient to overcome the resistance of the material. The energy release rate, G, is defined as the rate of change in potential energy with crack growing area for a linear elastic material. The material fracture occurs, when $G=G_c$, which is a measure of fracture toughness or critical energy release rate.

Damage tolerance entails allowing subcritical flaws to remain in a structure. Fracture mechanics provides a rational basis for establishing flaw tolerance limits and thus developing damage tolerant design. Consider a flaw in a structure that grows with time as schematically illustrated in Figure 3.7. The initial crack or flaw size may be inferred from a nondestructive examination (NDE), such as C-scan, ultra sound etc. The critical crack size may be computed or predicted from the applied stress or strain and fracture toughness. The allowable crack size can be defined via a safety factor or margin of safety factor. The life of the structure can then be estimated by calculating the time required for the flaw to grow from its initial size to the maximum allowable size. The other ingredients in this life estimation process are a crack growth law and knowledge of the stress environments the joint is subjected to. For adhesive bonded joint, a debond is a major crack or flaw. Debonding can occur at the termination of the bond line or within the bondline. For bonded composite joints, cracking or delamination may occur in the composite adherends in or near the overlap or joining section (see Figures 3.1 and 3.3).

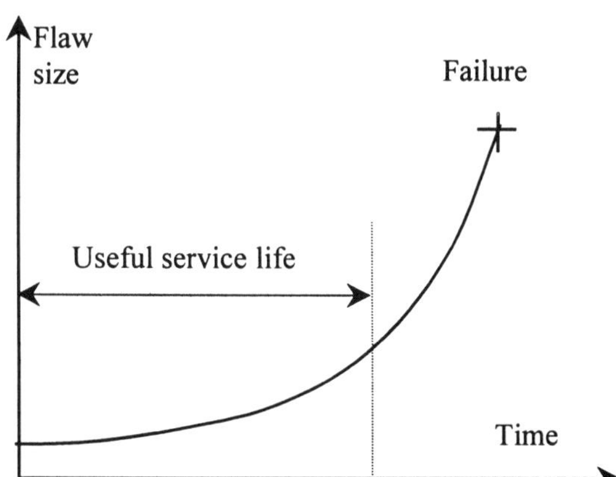

Figure 3.7 Damage tolerance approach to design

3.3.1 Energy release rate and stress intensity factor

(a) Energy release rate

Irwin (1956) defined an energy release rate, G, as the rate of change in potential energy with crack area A

$$G = -\frac{d\Pi}{dA} \tag{3.7}$$

Which is a measure of the energy available for an increment of crack extension. G is also called crack extension force or the crack driving force since it is obtained from the derivative of a potential. The potential energy of an elastic body, Π, is defined as follows:

$$\Pi = U - W \tag{3.8}$$

where U is the strain energy stored in the elastic body and W is the work done by external forces.

Crack propagation occurs when G attains a critical value G_C, namely, $G=G_C$. The crack growth may be stable or unstable, depending on how G and G_C vary with crack size. A plot of G_C versus crack propagation is referred to as a resistance curve or R-curve, while the G versus crack extension is called the driving force curve.

(b) Stress intensity factor

In terms of the coordinates shown in Figure 3.8, the local stress field in the vicinity of a crack in a sheet of linear elastic material is given by

$$\sigma_{ij} = \left(\frac{K}{\sqrt{\pi r}}\right) f_{ij}(\theta) \tag{3.9}$$

where K is called stress intensity factor, $f_{ij}(\theta)$ is a dimensionless function of θ. Stress intensity factor was due to the original work of Irvin and Westergaard. However, readers seeking more complete information are recommended to read the text by Anderson (1995).

Corresponding to the mode of loading, the stress field ahead of a crack tip in an isotropic linear elastic material can be written as

$$\lim_{r\to 0}\sigma_{ij}^{\ I} = \frac{K_I}{\sqrt{2\pi r}} f_{ij}^{\ I}(\theta) \tag{3.10a}$$

$$\lim_{r\to 0}\sigma_{ij}^{\ II} = \frac{K_{II}}{\sqrt{2\pi r}} f_{ij}^{\ II}(\theta) \tag{3.10b}$$

$$\lim_{r\to 0}\sigma_{ij}^{\ III} = \frac{K_{III}}{\sqrt{2\pi r}} f_{ij}^{\ III}(\theta) \tag{3.10c}$$

for the mode I, II and III, respectively (as shown in Figure 1.6).

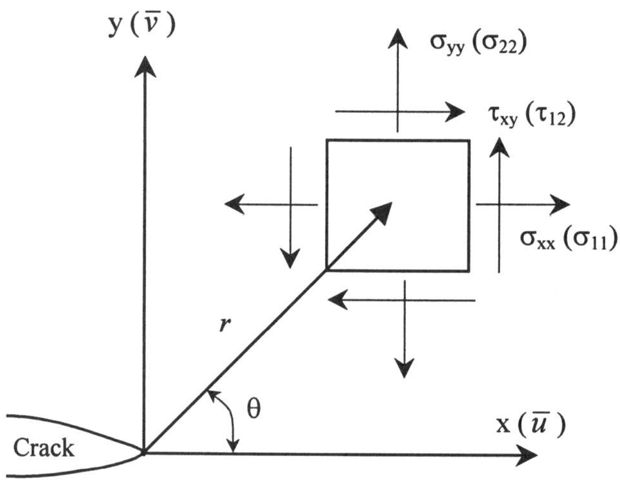

Figure 3.8 definition of the coordinate axis ahead of crack tip and stress components.

For a two-dimensional plane stress or strain case, the complex variable method has been widely used to present solutions for the stress and displacement fields near a crack tip. The complex stress field along the interface at a distance *r* ahead of a crack tip is given by

$$\sigma_{22} + i\sigma_{12} = \frac{K}{\sqrt{2\pi r}} \Gamma^{i\varepsilon} \tag{3.11a}$$

where $K = K_I + iK_{II}$ is the complex stress intensity factor.

The crack opening displacements at a distance r behind the crack tip are given by

$$\delta_2 + i\delta_1 = \frac{c^{(1)} + c^{(2)}}{2\sqrt{2\pi}(1+2i\varepsilon)\cosh(\pi\varepsilon)} \mathrm{K}\sqrt{r}r^{i\varepsilon} \tag{3.11b}$$

where the compliance parameters, $c^{(1)}$ and $c^{(2)}$, are

$$c^{(1)} = \frac{\kappa^{(1)}+1}{\mu^{(1)}}, \qquad c^{(2)} = \frac{\kappa^{(2)}+1}{\mu^{(2)}}$$

and ε is the oscillation index defined by the properties of the two materials

$$\varepsilon = \frac{1}{2\pi}\ln\left[\frac{\mu^{(1)} + \mu^{(2)}\kappa^{(1)}}{\mu^{(2)} + \mu^{(1)}\kappa^{(2)}}\right] \tag{3.12}$$

where μ is shear modulus and κ=3-4ν for plane strain and κ=(3-ν)/(1+ν) for plane stress. The above equations form the basis of the mode mixity analysis to be discussed in Section 3.3.5.

(c) The Relationship between *K* and *G*

The energy release rate quantifies the net change in potential energy with an increment of crack extension. The stress intensity factor characterizes the stresses, strains, and displacements in the vicinity of a crack tip. The energy release rate describes a global behavior, while the stress intensity factor is a local parameter. For linear elastic materials, these two parameters are uniquely related.

According to the Westergarrd's representations for the stresses and displacements in the vicinity of a crack, the relationships between the energy release rate and the stress intensity factor are

$$G_I = \frac{K_I^2}{E}\beta, \qquad G_{II} = \frac{K_{II}^{\ 2}}{E}\beta \tag{3.13}$$

where β equals to 1 for plane stress and 1-ν^2 for plane strain and E is Young's modulus.

3.3.2 J-integral

Linear elastic fracture mechanics (LEFM) is valid only when nonlinear material deformation is confined to a small zone around the crack tip. For many materials, LEFM is no longer applicable. Elastic-plastic fracture mechanics applies to materials that experience non-linear behavior, such as plastic deformation.

For linear elastic materials the energy release rate is defined by equation (3.7). The same definition can be used for nonlinear materials, except that G is replaced by J, namely

$$J = -\frac{d\Pi}{dA} \tag{3.14}$$

where A is crack area and Π is the potential energy given by equation (3.8). Thus, J is a more general version of the energy release rate. J=G holds for the special case of a linear elastic material.

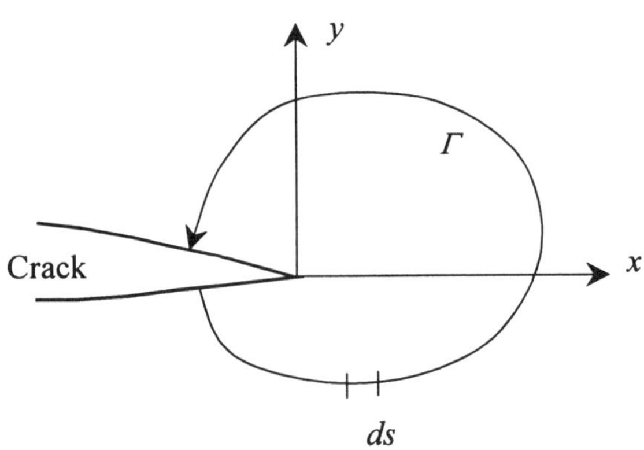

Figure 3.9 An arbitrary contour around a crack tip

Rice (1968) showed that the nonlinear energy release rate, J, could be written as a path-independent contour line integral. For an arbitrary counter-clock wise path, Γ, around a crack tip, as illustrated in Figure 3.9, the J integral is given by

$$J = \int_{\Gamma} (w dy - T_i \frac{\partial u_i}{\partial x} ds) \tag{3.15}$$

where w is the strain energy density, T_i are the traction components, u_i are the displacement components in the x and y direction respectively, and ds is a length increment along the contour Γ. The strain energy density is defined in equation (3.1). The traction is a stress vector normal to the contour, and its components are

$$T_i = \sigma_{ij} n_j \tag{3.16}$$

where n_j are the components of the unit vector normal to Γ.

Rice (1968) showed that the value of the J-integral is independent of the path of integration around the crack. Therefore J is called a path-independent integral. For linear elastic materials, it is relatively straightforward to determine the value of the J-integral in a test specimen or structure because J=G and G is uniquely related to the stress intensity factors. However, when the material is nonlinear, determination of the J integral becomes more difficult. Using finite element analysis method, stresses, strains and displacements along the path Γ can be calculated using the path-independent integral J value. This method can be used to both linear and nonlinear materials. In elastic materials, J can be evaluated at a remote contour, where numerical accuracy is better. In nonlinear materials that undergo path-dependent plastic deformations or thermal strains, J can still be computed at a remote contour with an appropriate correction term (Carpenter et al 1986)

3.3.3 Virtual crack closure integral method

According to Irwin (1958), if a crack extends by an infinitesimal amount Δ_c, the energy released in the process is equal to the work required to close the crack to its original length. Using the notation of Figure 3.8, the energy release rates for mode I and mode II deformations can be expressed as

$$G_I = \lim_{\Delta c \to 0} \frac{1}{2\Delta_c} \int_0^{\Delta c} \sigma_{yy}(\Delta c - r, 0) \overline{v}(r, \pi) dr \tag{3.17a}$$

$$G_{II} = \lim_{\Delta c \to 0} \frac{1}{2\Delta_c} \int_0^{\Delta c} \tau_{xy}(\Delta c - r, 0)\overline{u}(r, \pi)dr \tag{3.17b}$$

where G_I and G_{II} are the energy release rates for mode I and II deformation, σ_{yy} and τ_{xy} are the stresses ahead of the crack tip, and $\overline{u}$ and $\overline{v}$ are the displacements behind the crack tip.

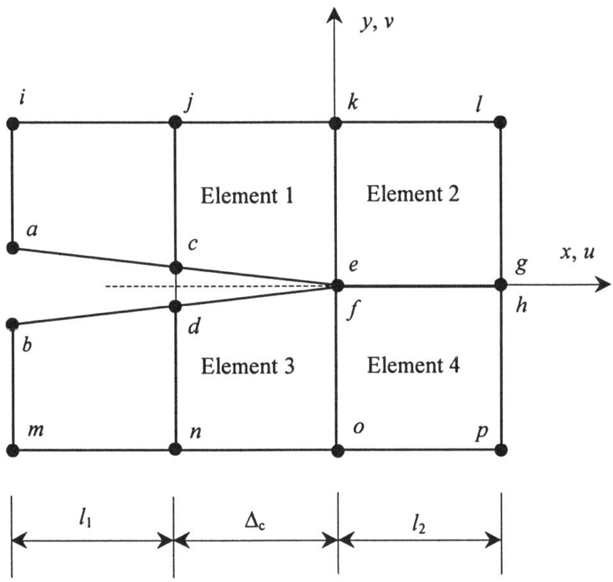

Figure 3.10 The finite elements and nodes near crack tip

Rybicki and Kanninen (1977) proposed a procedure to obtain these energy release rates for a homogeneous isotropic material from finite element solutions. In their procedure, the energy release rates of mode I and mode II deformations can be evaluated via the nodal forces and displacements with four-node isoparametric finite elements as shown in Figure 3.10 using the following expressions

$$G_I = \lim_{\Delta c \to 0} \frac{1}{2\Delta_c} \overline{F}_{yc}(v_c - v_d) \tag{3.18a}$$

$$G_{II} = \lim_{\Delta c \to o} \frac{1}{2\Delta_c} \overline{F}_{xc}(u_c - u_d) \tag{3.18b}$$

where Δ_c is the size of the element at the crack tip, $\overline{F}_{yc}$ and $\overline{F}_{xc}$ are the nodal forces in the y and x directions, respectively, that are required to hold node c and d together. They also recommend that l_1 and l_2 should be maintained equal to Δ_c.

Raju (1987) employed higher order and singular finite elements in order to improve the accuracy of the computed energy release rates of mode I and mode II deformations.

The virtual crack closure integral is an accurate and efficient method in computing the mixed-mode energy release rate for a crack in a homogeneous isotropic material. However when applied to an interfacial crack between two dissimilar isotropic and anisotropic materials, it was found that the individual components, mode I and mode II, of the energy release rates computed from finite element model are strongly mesh dependent (Sun and Jih, 1987; O'Brien, 1982). Raju et al (1988) showed that the mixed-mode energy release rates are functions of $\Delta^{i\varepsilon}$ where Δ is the length of the crack assumed to be closed and ε is the oscillation index defined in equation (3.12). In the finite element analysis, Δ is also the mesh size of the elements around the crack tip, therefore the calculated mixed-mode energy release rates depends on the mesh refinement near the crack tip.

To resolve this problem, the following coupled energy release rate $G_{I\text{-}II}$ was introduced (Atluri, 1997)

$$G_{I-II} = \frac{1}{2\Delta c} \int_0^{\Delta c} \left[\sigma_{12}(r)\delta_2(\Delta_c - r) + \sigma_{22}(r)\delta_1(\Delta_c - r)\right] dr \tag{3.19}$$

Unlike the mixed-mode energy release rates, the mixed-mode stress intensity factors of an interfacial crack between dissimilar isotropic media, first defined by Williams (1959), and Rice and Sih (1965), are independent of the crack closing length. The mixed-mode stress intensity factors are hence preferred by many researchers as the fracture parameters to define the toughness criterion of an interfacial crack. Atluri (1997) employed the virtual crack closure integral approach and proposed simple method for computing the stress intensity factors for an interfacial crack. Using the crack opening displacements and the nodal forces at and ahead of the crack tip, the stress intensity factors can be calculated for an isotropic bimaterial continuum, an orthotropic bimaterial continuum, and an anisotropic bimaterial continuum.

3.3.4 Mixed-mode fracture and failure criteria

Propagation of a crack in the bondline is a major failure mechanism of adhesive joints. Thus fracture mechanics has been widely used to study the crack-growth behavior in adhesive joints. A crack in the bondline of an adhesive joint is constrained by the stiff adherends, and propagates in general under a mixed-mode deformation. There are several test methods for characterizing mode I, mode II and mixed-mode fracture toughness of adhesive joints as described in Chapter 1. For example, the double cantilever beam (DCB) test are often used for measuring the mode I fracture toughness and the end notch flexure (ENF) test for mode II fracture toughness. In addition, there are a number of tests for measuring mixed-mode fracture toughness. These include the cracked lap shear test (Johnson, 1987), the mixed-mode-flexure test (Russel and Street, 1982), the asymmetric-double-cantilever-beam test (Bradley and Cohen, 1985), the fixed ratio mixed mode test, the varying ratio mixed mode test (Hashemi et al, 1991), and the mixed-mode bending test (Reeder and Crews, 1990). Ferlund and Spelt (1994a,b) also proposed a test allowing fracture testing in the entire mode-ratio range from pure mode I to pure mode II.

A failure criterion can be postulated as: bond or adhesive failure in adhesive bonded lap joints occurs when the maximum strain energy release rate in the adhesive layer attains its ultimate value G_C for a mixed-mode deformation, namely,

$$G(\psi) = G_C(\psi) \tag{3.20}$$

Where $G_C(\psi)$ is the adhesive failure envelope that is defined as the critical strain energy release rate of an adhesive subject to a mixed-mode I and II deformation. Although it is desirable to measure the entire in-situ failure envelope for a specific material combination, measurement of such in-situ failure envelope requires enormous effort. As it is only applicable to one particular material combination, such measurement can be very expensive. For this reason, simple and interactive fracture failure criteria have been developed (Chai, 1993; Reeder, 1993).

A general form of such interactive failure criterion of mixed-mode I and II fracture can be written as follows:

$$\left(\frac{G_{II}}{G_{IIC}}\right)^{\alpha}+\left(\frac{G_{I}}{G_{IC}}\right)^{\beta}=1 \qquad (3.21$$

where α and β are real constants, and G_{IC} and G_{IIC} are the critical bond strain energy rates of pure peel and pure shear respectively. In the following section, we will evaluate two simple cases:

Linear failure criterion ($\alpha=\beta=1$):

$$\frac{G_{II}}{G_{IIC}}+\frac{G_{I}}{G_{IC}}=1 \qquad (3.22)$$

Quadratic failure criterion ($\alpha=\beta=2$):

$$\left(\frac{G_{II}}{G_{IIC}}\right)^{2}+\left(\frac{G_{I}}{G_{IC}}\right)^{2}=1 \qquad (3.23)$$

For cracks in brittle and homogeneous materials, many efforts have been focused on cracks under mode I loading, since there is ample experimental evidence that a crack under mixed-mode loading will grow by kinking in a irection such that the crack tip is in pure mode I. Figure 3.11 depicts the widely used tapered double cantilever beam (TDCB) test specimen, for example used by Ripling et al (1971, 1983, 1988) and Mostovoy and Ripling (1975) to characterize the fracture behavior under mixed-mode static and cyclic loading.

For the TDCB joint, the critical strain energy release rate G_C may be determined using the LEFM approach via the following the formula

$$G_C=\frac{P_C^{\,2}}{2b}\frac{dC}{da} \qquad (3.24)$$

where P_C is the fracture load, C is the compliance constant and $C=\delta/P$, in which δ is the displacement.

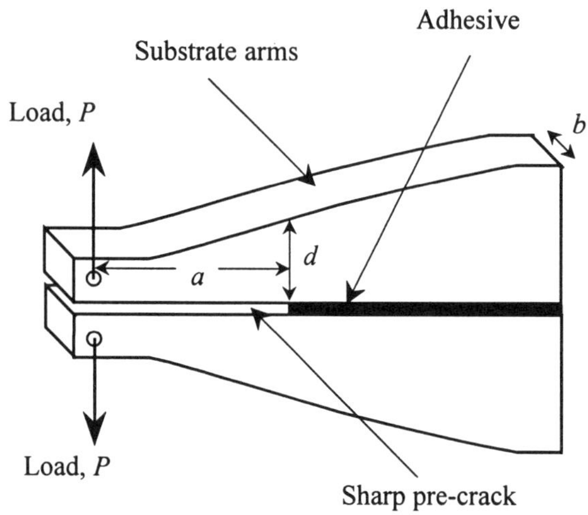

Figure 3.11 A tapered double cantilever-beam (TDCB) joint specimen under an opening mode loading (mode I).

For a homogeneous beam with a thin adhesive layer (i.e. neglecting contribution of adhesive layer), the compliance change per unit crack growth, dC/da, can be calculated viz. (Kinloch, 1987)

$$\frac{dC}{da} = \frac{8}{E_S bd}\left(1 + \frac{3a^2}{d^2}\right) \tag{3.25}$$

where E_S is the modulus of the substrate arms. By combining equations (3.24) and (3.25), the critical strain energy release rate can be determined using

$$G_C = \frac{4P_C^{\ 2}}{E_S db^2}\left(1 + \frac{3a^2}{d^2}\right) \tag{3.26}$$

Table 3.1 tabulates some measured critical energy release rates G_{IC} and G_{IIC} for some adhesive and adherend combinations from various sources.

As shown in the table, for the case of adherend made of the same material 7075-T6 Aluminum, the values of critical energy release rates of mode I and II for the epoxy Cybond 4523GB adhesive are much lower than those for the Permabond ESP 310 adhesive. This comparison early indicates that the Permabond ESP 310 adhesive is tougher than the Cybond 4523GB adhesive when used to bond 7075-T6 aluminum. It is also seen that the mode I fracture toughness of Permabond ESP 310 adhesive is larger than that of FM 300K film adhesive when sandwiched in similar aluminum alloys.

Table 3.1 Measured critical energy release rates for some material combinations (for reference only)

Adhesives and manufacture	Adherend materials	G_{IC} (J/m^2)	G_{IIC} (J/m^2)
Epoxy Cybond 4523GB (American Cynamaid)[1]	7075-T6 Aluminum	212	575
Permabond ESP 310 (American Cynamaid)[2]	7075-T6 aluminum	794	5605
FM 300K .05 (American Cynamaid) [3]	7075-T651 aluminum alloy	654	-
Ferokal 4520-34 (Teroson)[4]	EG steel	740	-
XD 4600 (Ciba Polymers)[4]	Aluminum-alloy	3500	-

[1] Ferlund et al (1994); [2] Papini et al (1994); [3] Ripling et al (1988); [4] Jethwa and Kincloch (1997).

Figure 3.12 compares the measured and the predicted energy release rate J_C for 7075-T6 aluminum adherends bonded with Cybond 4523GB epoxy adhesive system (the measured curve taken from Fernlund et al (1994). The predicted failure envelopes are calculated using equations (3.22) and (3.23) with J_{IC}= 212 J/m^2 and J_{IIC}=575 J/m^2. It is seen that J_C given by the quadratic criterion is always larger than the measured values, while J_C computed using the linear criterion is larger when the phase angle is less than about 43°, and is less when the phase angle is larger than 43°. For the linear criterion, the relative differences are between -20% and 25%, while for the quadratic criterion, the relative difference can be as high as 40%. The linear criterion appears to be more appropriate when the phase angle is larger than 34°. It is noted that the quadratic criterion over predicts J_C and thus can result in an unconservative failure prediction.

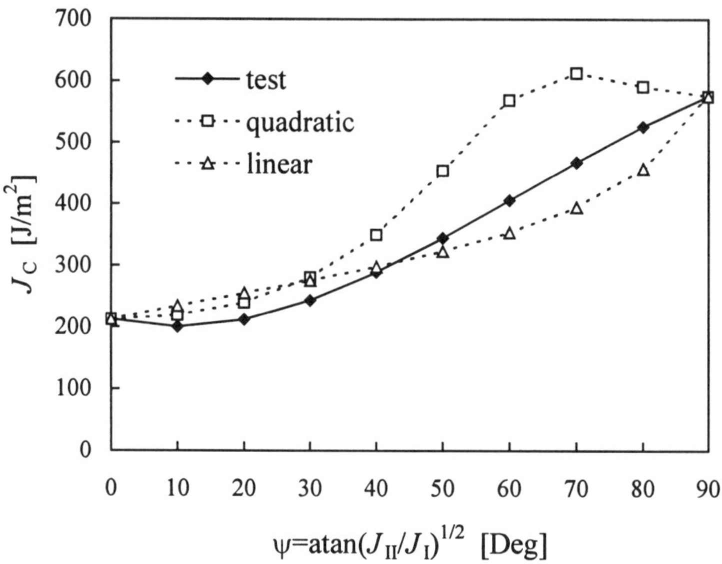

Figure 3.12 Failure envelopes, J_c versus phase angle ψ (Tong, 1998a)

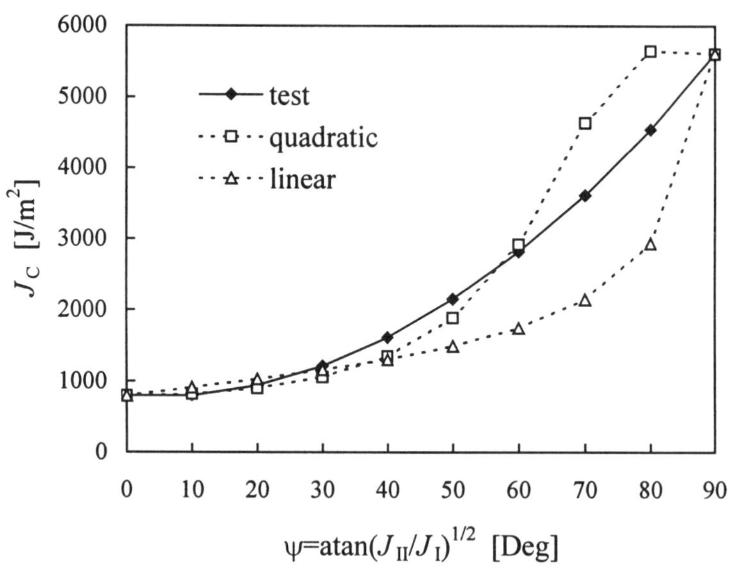

Figure 3.13 Failure envelopes, J_c versus phase angle ψ (Tong, 1998a)

Figure 3.13 plots the measured and the predicted energy release rate J_C for 7075-T6 aluminum adherends bonded with Permabond ESP 310 adhesive system (Test curve taken from Papini et al 1994). Once again, equations (3.22) and (3.23) are used to calculate the value of J_C with J_{IC}=794 J/m^2 and J_{IIC}=5605 J/m^2. There is little difference between the quadratic criterion and the linear criterion when the phase angle is less than 40°. When the phase angle is greater than 40°, the quadratic criterion tends to be more appropriate indicator than the linear criterion as it is more close to the measured curve.

It is noted that the present two simple failure criteria given in equation (3.22) and (3.23) can be used to generate approximate failure envelopes. In this case, only the critical values of the strain energy release rates in pure mode I and mode II need to be measured.

3.3.5 Calculation of mixed-mode fracture toughness

The apparent fracture toughness of an adhesive joint is in general higher than that of a bulk adhesive because adhesives are usually less rigid than adherends and the elastic mismatch of the adhesive and adherends causes a constraining effect.

Adhesives in adhesive joints may exhibit brittle or ductile behavior. In a brittle adhesive, deformation is confined to linear proportional limit or if the size of the plastic zone is small compared with adhesive thickness. In a ductile adhesive, a large plastic zone much greater than the adhesive thickness (Chai, 1993b). Fleck et al (1991) suggested that an adhesive layer is brittle if the size of the plastic zone around the crack tip, τ_p, is small compared with the layer thickness, H,

$$\tau_p \cong \frac{1}{3\pi}\left(\frac{\mathrm{K}_{IC}}{\sigma_y}\right)^2 < \frac{H}{25}$$

where K_{IC} is the fracture toughness and σ_y is the yield stress of the bulk adhesive, respectively.

Experimental results show that a crack propagates continuously in and isotropic homogeneous brittle solid following a path where K_{II}=0. An adhesive layer subject to remote pure mode I loading is thus under pure mode I locally. When a mode II loading is added to the joint, the crack then find a pure mode I path off the centerline.

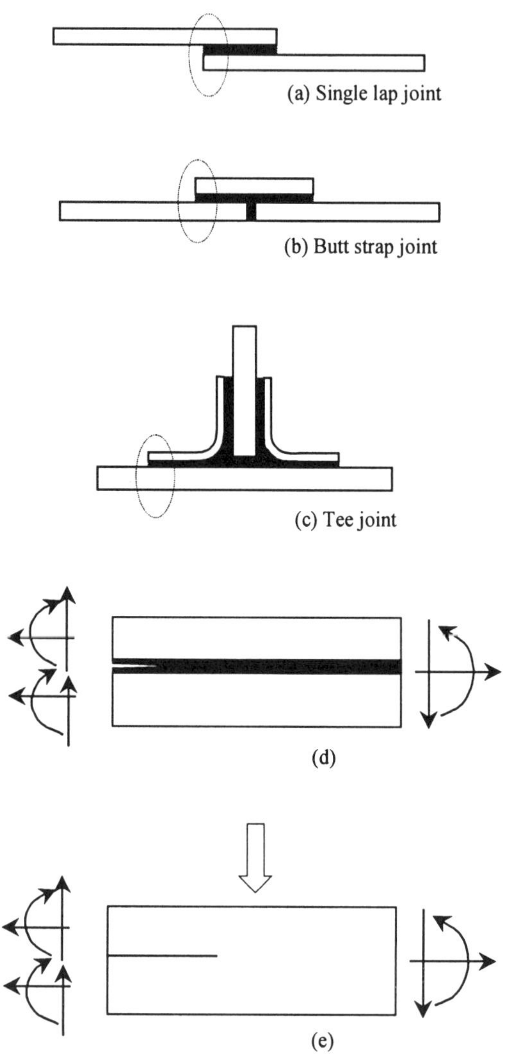

Figure 3.14 Mechanics model for bonded joints with a crack in the adhesive

It is a common practice to calculate the fracture energy G_C for adhesive joints from the measured critical loads by ignoring the presence of the adhesive layer. As shown in Figure 3.14a-c, a crack may develop in adhesive or along the adhesive-adherend interface near an adhesive end in various joints. The portion containing the crack can also be represented by a free body diagram, shown in Figure 3.14d, once all the stress resultants are known.

When applying the various fracture mechanics method to calculate the fracture parameters, such as energy release rates and stress intensity factors, it is assumed that the thickness of adhesive layer is very small compared with the dimensions of the joint adherends. Thus the fracture parameters for adhesive joints are global quantities.

Using a simple energy argument or the J-integral, the global energy release rate can be proved to be equal to the local energy release rate at the crack tip. However, the global stress intensity factors are not always equal to the local stress intensity factors at the crack tip, and there exist a relationship between the global and local stress intensity factors (Fleck et al (1991)).

Figure 3.14(e) shows a model that neglects the presence of the adhesive layer. This is a valid practice when the adhesive layer is thin and the crack is long and parallel to the adhesive and adherend interface. According to the conservation of the J-integral, the calculated G_C is equal to the actual energy released at the crack tip regardless of the crack position, within the adhesive or along the interface. It should be kept in mind that the value of the fracture energy G_C is the cohesive fracture energy or the adhesive/adherend interface fracture energy at the relevant mode of loading depending whether the crack advances within the adhesive layer or along the interface.

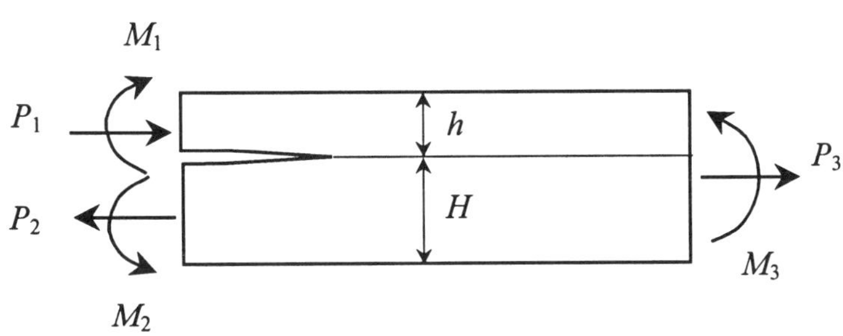

Figure 3.15 A crack lies in an isotropic material

To predict the fracture load of an adhesive bonded joint using a fracture mechanics based approach, it is important to determine the energy release rate of mode I and II at the crack tip as shown in Figure 3.14(e). Williams (1988) presented an analysis for determining the energy release rates of mode I and II for cracked laminates. For the problem of mixed-mode cracking in layered materials, Hutchinson and Suo (1992) summarized their complete solutions for the following two case: (a) a crack lies in an isotropic material (see Figure 3.15), and (b) a crack lies on the bimaterial interface in a composite laminate (see Figure 3.16).

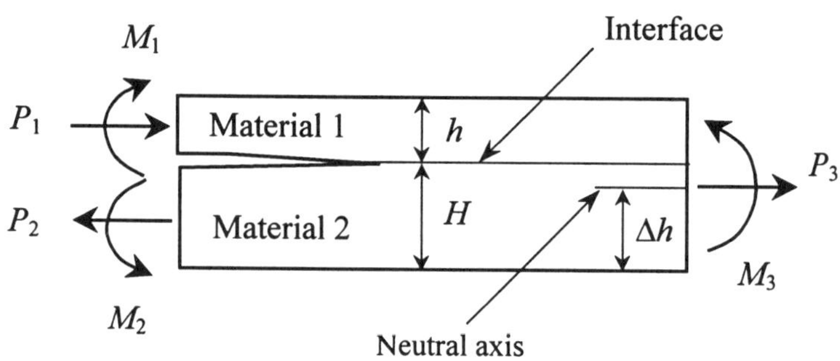

Figure 3.16 A crack lies on an bimaterial interface

(a) Mode mixity for a crack in an isotropic material
Total energy release rate at the crack tip shown in Figure 3.15 can be calculated using (Hutchinson and Suo, 1992)

$$G = \frac{1}{2}\left[\frac{P_1^{\,2}}{\overline{E}h} + \frac{P_2^{\,2}}{\overline{E}H} + \frac{12M_1^{\,2}}{\overline{E}h^3} + \frac{12M_2^{\,2}}{\overline{E}H^3} - \frac{P_3^{\,2}}{\overline{E}(h+H)} - \frac{12M_3^{\,2}}{\overline{E}(H+h)^3}\right] \tag{3.27}$$

and the mode I and II components of the energy release rates are

$$G_I = \frac{1}{2\overline{E}}\left[\frac{P}{\sqrt{hU}}\cos\omega + \frac{M}{\sqrt{h^3V}}\sin(\omega+\gamma)\right]^2 \tag{3.28a}$$

$$G_{II} = \frac{1}{2\overline{E}}\left[\frac{P}{\sqrt{hU}}\sin\omega - \frac{M}{\sqrt{h^3V}}\cos(\omega+\gamma)\right]^2 \tag{3.28b}$$

where $\overline{E}$ is the modulus, and

$$P = P_1 - C_1P_3 - C_2M_3/h, \qquad M = M_1 - C_3M_3$$

$$C_1 = \frac{1}{1/\eta+1}, \quad C_2 = \frac{6/\eta}{(1/\eta+1)^3}, \qquad C_3 = \frac{1}{(1/\eta+1)^3}$$

$$\eta = h/H, \qquad \frac{1}{U} = 1 + 4\eta + 6\eta^2 + 3\eta^3$$

$$\frac{1}{V} = 12(1+\eta^3), \qquad \frac{\sin\gamma}{\sqrt{UV}} = 6\eta^2(1+\eta) \tag{3.29}$$

and ω varies slowly with η in the entire range $0 \le \eta \le 1$, and can be approximated as

$$\omega = 52.1° - 3°\eta \tag{3.30}$$

(b) Mode mixity for a crack on an bimaterial interface

Using the J integral, the total energy release rate at the crack tip shown in 3.16 can be calculated in the following close form (Hutchinson and Suo, 1992):

$$G = \frac{1}{2\overline{E}_1}\left(\frac{P_1^2}{h} + 12\frac{M_1^2}{h^3}\right) + \frac{1}{2\overline{E}_2}\left(\frac{P_2^2}{H} + 12\frac{M_2^2}{H^3} - \frac{P_3^2}{Ah} - \frac{M_3^2}{Ih^3}\right) \tag{3.32}$$

This total energy release rate gives only the magnitude of the near-tip singularity but not mode mixity. The mode mixity can be given in the stress intensity factor K form as follows

$$K = h^{-i\varepsilon}\left(\frac{1-\alpha}{1-\beta^2}\right)^{1/2}\left(\frac{P}{\sqrt{2hU}} - ie^{i\gamma}\frac{M}{\sqrt{2h^3V}}\right)e^{i\omega} \tag{3.33}$$

Where $i = \sqrt{-1}$, and P and M are linear combinations of the edge loads:

$$P = P_1 - C_1P_3 - C_2M_3/h, \qquad M = M_1 - C_3M_3 \tag{3.34}$$

and

$$C_1 = \frac{\Sigma}{A}, \qquad C_2 = \frac{\Sigma}{I}\left(\frac{1}{\eta} + \frac{1}{2} - \Delta\right), \; C_3 = \frac{\Sigma}{12I}$$

$$\frac{1}{U} = 1 + \Sigma\eta\left(4 + 6\eta + 3\eta^2\right), \qquad \frac{1}{V} = 12\left(1 + \Sigma\eta^3\right)$$

$$\frac{\sin\gamma}{\sqrt{UV}} = 6\Sigma\eta^2(1+\eta), \;\; A = \frac{1}{\eta} + \Sigma,$$

$$I = \left[\left(\Delta - \frac{1}{\eta}\right)^2 - \left(\Delta - \frac{1}{\eta}\right) + \frac{1}{3}\right] + \frac{\Delta}{\eta}\left(\Delta - \frac{1}{\eta}\right) + \frac{1}{3\eta^3}$$

$$\Delta = \frac{1 + 2\Sigma\eta + \Sigma\eta^2}{2\eta(1+\Sigma\eta)}, \qquad \Sigma \equiv \frac{\overline{E}_1}{\overline{E}_2} = \frac{1+\alpha}{1-\alpha}, \qquad \eta = \frac{h}{H}$$

$$\alpha = \overline{E}_1 - \overline{E}_2/\left(\overline{E}_1 + \overline{E}_2\right),$$

$$\beta = \frac{1}{2}\frac{\mu_1(1-2\upsilon_2) - \mu_2(1-2\upsilon_1)}{\mu_1(1-\upsilon_2) + \mu_2(1-\upsilon_1)} \tag{3.35}$$

The mode mixity can be determined using equation (3.32-33) and the following equations

$$G = \frac{1-\beta^2}{2}\left(\frac{1}{\overline{E}_1} + \frac{1}{\overline{E}_2}\right)\left(\mathrm{K}_I^{\,2} + \mathrm{K}_{II}^{\,2}\right) \tag{3.36}$$

3.3.6 Two Illustrative examples

An ASTM round-robin effort was made to address the problem of calculating the strain energy release rate for the cracked lap shear (CLS) specimen (Johnson, 1987). The CLS specimen was chosen because it has been used for studying composite delamination and adhesive joint debonding. Figure 3.17 shows the cracked lap-shear joint specimen. The joint specimen has a width of 25.4 mm. The lap adherend length is 254 mm, and the strap adherend is 305 mm long. The adhesive bondline thickness is 0.13 mm. Two cases were considered, one assumed that both adherends have the same thickness of 6.35 mm, and the other assumed that the lap and strap adherends are 12.7 and 6.35 mm thick respectively.

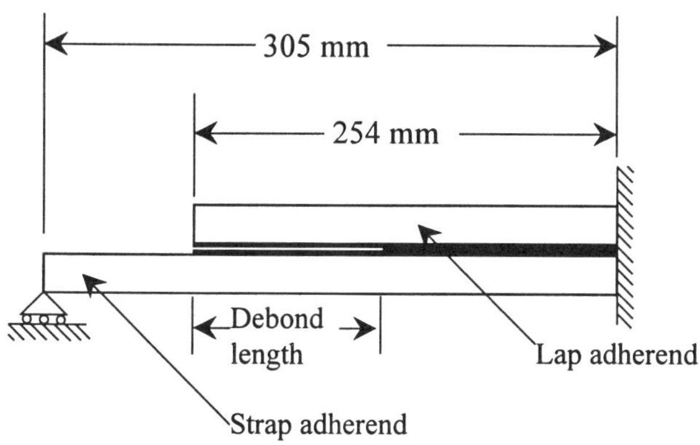

Figure 3.17 Cracked lap shear specimen of equal and unequal thickness

The adherends were assumed to be aluminum with an elastic modulus of 72.45 GPa and a Poisson's ratio of 0.33. The adhesive was assumed to be linear elastic and has a modulus of 1.932 GPa and a Poisson's ratio of 0.40. A tensile load of 11.12 KN was applied to both specimen geometries. the debond was assumed to be located at the middle of the adhesive layer, representing an unusual case in practice: cohesive crack propagation. The analysis procedures used by various groups fall into three general classes: closed form, finite element analysis, and mixed closed form/finite element analysis. In the finite element analysis, geometrical linear and nonlinear cases were also considered separately and the virtual crack closure method or a modified version was used to calculate the strain-energy density. The geometric nonlinear analyses result in rather constant total energy release rate G_T and same G_I/G_{II} ratio for the equal and unequal adherends specimens in spite of the rather significant charge in debond length. The geometric nonlinear FEA results are most consistent with observed experimental behavior, and thus the nonlinear analysis should be used for analyzing the CLS specimen. The closed-form approach based on the classical beam theory by Brussat et al (1977) gives a good approximation of G_T for both specimens, and also a good approximation of G_I/G_{II} ratio only for specimen of same thickness adherends.

The cracked lap shear specimen was revisited by Lai et al (1996). A new non-dimensional closed form solution was introduced based on a theory similar to that proposed by Goland and Reisser (1994) for a classical single lap joint problem. In their analysis, the strap adherend and the lap adherend were assumed to have different thickness and to be made of different materials. They consider the effect of various boundary condition, namely, clamped-clamped, roller-clamped and roller-roller. Using these various boundary conditions at both end, the bending moment was determined. The fracture parameters were then determined using the energy release rate and the stress intensity factors (Suo and Ilutchinson, 1992) for an interfacial crack in a bilayer with each layer being isotropic, homogeneous and linearly elastic. They also provide a comparison between their closed-form solutions and numerous results of nonlinear finite element analysis taken from the literature.

3.3.6 Relationship between energy release rate and the strain energy density

For a balanced adhesive-adherend sandwich subjected to a combination of loads at its end, Tong (1996) demonstrates that the energy release rate J is equal to the product of the strain energy density W as defined in equation (3.1) and the adhesive thickness t, namely,

$$J = tW \tag{3.37}$$

and

$$J_I = tW_I, \qquad J_{II} = tW_{II} \tag{3.38}$$

These equations are also true for the critical values of the energy release rates and the strain energy density, i.e.

$$t\int_0^{\gamma_{max}} \tau(\gamma)dr = J_{IIC} \qquad t\int_0^{\varepsilon_{max}} \sigma(\varepsilon)d\varepsilon = J_{IC} \tag{3.39}$$

This conclusion was found to be true for the cases of linear stress-strain relations by Edde and Verreman (1992) and for nonlinear shear behavior of adhesive by Chai (1993a). It should be borne in mind that the above conclusion is applicable only when the adhesive layer is thin and the shear strain in the adhesive layer can be approximated as constant across the adhesive thickness. These equations can serve as a cross-check of the measured values of the critical energy release rates and the stress-strain curves.

3.4 APPLICATIONS OF STRENGTH OF MATERIALS CRITERIA

In this section, illustrative examples will be presented to demonstrate the applications of the strength of materials based failure criteria in predicting the static strength of metallic and composite lap joints of various geometrical configurations. These examples include (a) prediction of static strengths of adhesive bonded composite double lap joints and panel-flange joints using the point-based failure criteria, and (b) prediction of static failure loads of the aluminum joints and composite joints for the zone-based failure criteria.

3.4.1 Examples using point-based criteria

(a) Composite double lap joints

Quasi static tensile tests were conducted for double lap joint specimens as schematically shown in Figure 3.18 with aspect ratio l/t being kept as 30 and 20 respectively. When the aspect ratio remained constant, overlap length varied with adherend thickness. The material used was the T650/F584 graphite/epoxy prepreg and FM-300K film adhesive. Metal shims were used to maintain a constant bond line thickness of 0.15 mm. Joint specimens of 25.4 mm wide were cut from the large bonded panels.

Table 3.2 tabulates the adherend lay-ups, adherend thickness, and the overlap length for all specimens. The mechanical properties for a typical ply are listed in Table 3.3 (Tong, 1997). The tensile and shear stress-strain curves for the adhesive are depicted in Figure 3.19. The tensile stress-strain curve was measured using the neat adhesive tensile specimens, while the shear-strain curve was measured using the thick adherend lap shear test.

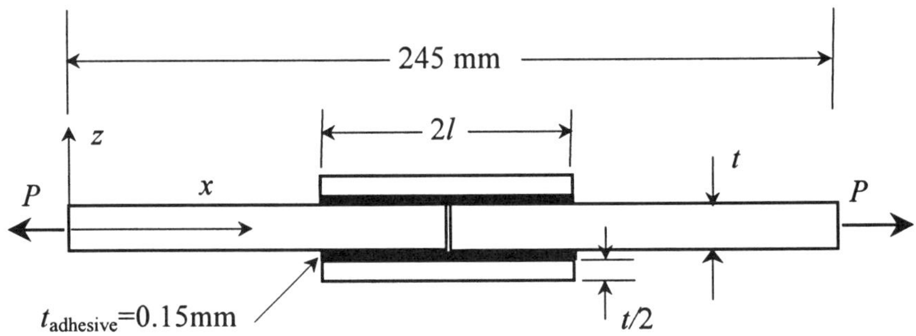

Figure 3.18 Geometric configurations of the balanced double lap joint specimen (redraw from Tong, 1997)

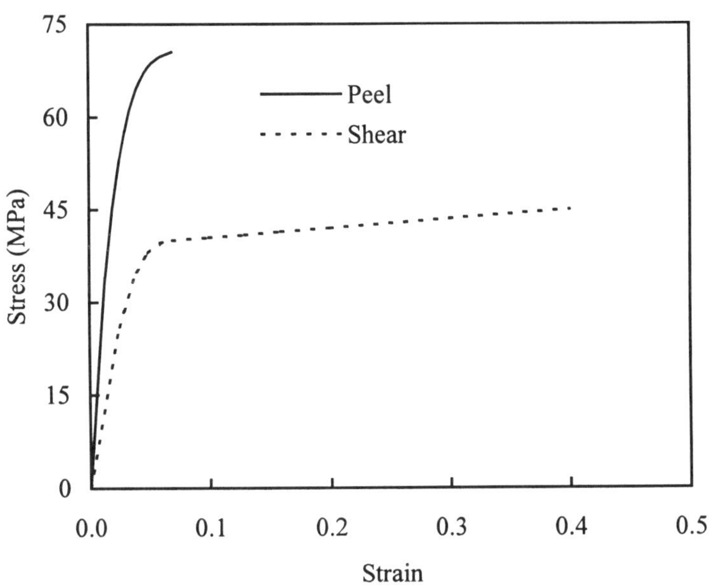

Figure 3.19 Stress-strain curves for FM300-K film adhesive

Table 3.2 Parameters for balanced double lap joint specimens with l/t of 30 and 20 (dimension in mm)

Specimen Group	Doubler Lay-up	Central lay-up	Adherend Thickens t	Overlap length l
A2	[0,90]s	[0,90]2s	1.12	33.5
B2	$[0_2,90]$s	$[0_2,90]$2s	1.67	50.1
C2	$[0_2,90_2]$s	$[0_2,90_2]$2s	2.23	67.1
A3	[0,90]s	[0,90]2s	1.12	22.4
B3	$[0_2,90]$s	$[0_2,90]$2s	1.67	33.4
C3	$[0_2,90_2]$s	$[0_2,90_2]$2s	2.23	44.6

Table 3.3 Typical material properties of a T650/F584 graphite/epoxy lamina (nominal ply thickness is 0.14 mm)

Longitudinal modulus E_L	156.5	GPa
Transverse Modulus E_T and E_Z	15.65	GPa
Poisson ratio ν_{LT} and ν_{LZ}	0.324	
Poisson ratio ν_{Tz} (neat resin)	0.35	
Shear modulus G_{LT} and G_{LZ}	5.19	GPa
Shear modulus G_{TZ} (neat resin)	1.53	GPa
Longitudinal tensile strength X_T	2337	MPa
Longitudinal compressive strength X_C	1585	MPa
Transverse tensile strength Y_T and Z_T	46.6	MPa
Interlaminar shear strength R and S	126.9	MPa

All specimens were loaded at a constant crosshead speed of 1 mm/min in an ambient environment. Inspection of the failure surfaces shows that the typical failure loci are near the adhesive ends and within the surface ply of the central adherend as shown in Figure 3.20a. Interlaminar cracking typically occurs near the top of the surface ply and then either propagates in that ply or kinks into the next ply, with broken fibers being observed similar to the phenomenon by Kairouz and Matthews (1993). Figure 3.20b depicts the typical stresses acting on an infinitesimal element taken from the critical areas marked by the circles in Figure 3.20a. Amongst these three stresses, the normal tensile stress σ_z is believed to be the predominant one that causes delamination, whereas the shear stress τ_{xz} and the axial tensile stress σ_x contribute to delamination through interacting with the normal tensile stress σ_z.

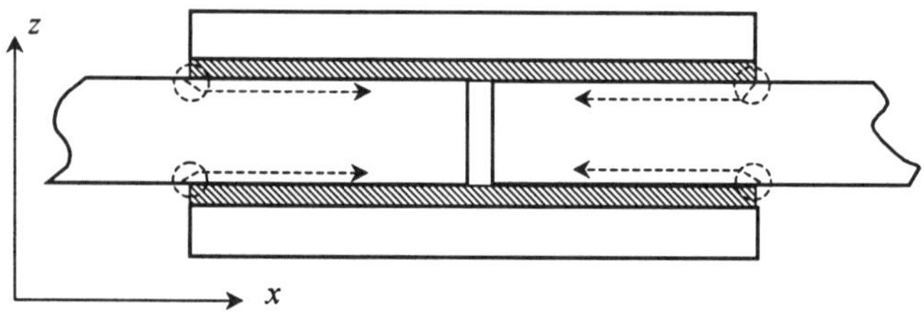

(a) A double lap joint

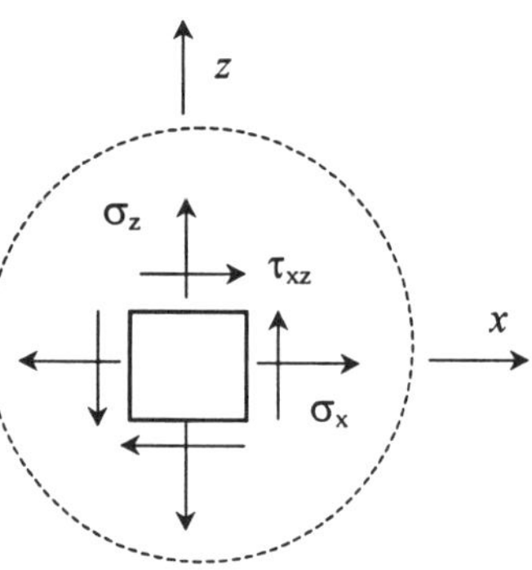

(b) Typical stresses in the circled failure loci

Figure 3.20 Typical failure loci (denoted by hollow circles), crack path (denoted by dotted lines) and the relevant stresses for a double lap joint specimen (redraw from Tong, 1997)

The minimum, maximum and average values of the measured failure loads per unit width are summarised in Table 3.4(a) for all specimens. Specimens in group A2 and A3 had the smallest failure loads because they had the thinnest adherends and hence the weakest membrane strength. Specimens in group B2 and B3 had the largest failure loads, and specimens in group C2 and C3 failed at the intermediate load levels. It is also seen that joints with thin outer adherends have higher joint efficiency (Tong, 1997).

Table 3.4(a) Measured failure loads (kN/m) for double lap joint specimens

Specimen	A2	B2	C2
Min	691.0	1241.7	1147.0
Max	860.1	1522.2	1416.2
Average	781.6	1401.6	1264.0
Av. joint efficiency	59.7%	53.5%	48.3%
Specimen	A3	B3	C3
Min	761.5	1280.6	1222.8
Max	951.5	1470.6	1405.9
Average	864.8	1393.6	1357.1
Av. joint efficiency	66.1%	53.2%	51.8%

* Average joint efficiency equals to the ratio of the maximum stress outside the joint to its pertinent strength allowable.

The adhesive and adherend sandwich model described in Section 2.3 is used to perform one-dimensional analysis. Calculated are the stresses in the adhesive as well as stresses in the adherends in the test specimens. The adhesive is modeled as being linear and nonlinear. In the nonlinear analysis, true tensile and shear stress-strain curves of figure 3.19 are used as the adhesive material data. In the linear elastic analysis, the Young's modulus and shear modulus are taken as the initial slopes of both curves. The failure criteria listed in equations (3.6.1-6) are utilized to predict joint strength with the peel and shear stresses in the adhesive layer and the axial stress in the surface ply.

A nonlinear plane-strain finite element model was developed to analyze the stresses and strains in the joints for the specimen group A2, B2 and C2 using the MSC/NASTRAN (1992). Large deformations are considered using the available geometrical nonlinear facilities. Composite adherends are modeled as being linearly elastic and using ply-by-ply meshing scheme in which a fine mesh is near the end of adhesive layer (Tong, 1997). The tensile stress-strain curve for FM300-K adhesive depicted in Figure 3.19 is used in conjunction with the Von Mises yielding criterion to model the nonlinear behaviour of the adhesive layer.

Table 3.4b tabulates the failure loads per unit width predicted using the six failure criteria in equation (3.6.1-6) for the following four cases:

(a) one-dimensional predictions using linear adhesive model,
(b) one-dimensional predictions utilizing nonlinear adhesive material properties,

(c) finite element predictions using the stresses in the center of the surface ply element, and

(d) finite element predictions employing the stresses in the center of the adhesive layer element.

Table 3.4(b) Predicted failure loads (kN/m) for double lap joints

One dimensional analysis results			
Specimen	A2 & A3	B2 & B3	C2 & C3
Predicted failure load using nonlinear adhesive properties			
Criterion 3.6.1	1054.9	1798.4	1716.2
Criterion 3.6.2	807.9	1339.3	1248.6
Criterion 3.6.3	1047.0	1782.1	1700.8
Criterion 3.6.4	800.2	1323.8	1230.1
Criterion 3.6.5	886.3	1521.1	1471.3
Criterion 3.6.6	671.0	1158.3	1050.1
Predicted failure load using linear elastic adhesive properties			
Criterion 3.6.1	528.8	717.2	712.1
Criterion 3.6.2	470.4	648.3	648.0
Criterion 3.6.3	527.3	715.7	710.6
Criterion 3.6.4	468.7	646.5	646.2
Criterion 3.6.5	513.7	706.9	702.9
Criterion 3.6.6	454.2	636.1	636.7
Finite element analysis results (surface ply element)			
Specimen	A2	B2	C2
Criterion 3.6.1	1100.2	1865.6	1785.4
Criterion 3.6.2	835.1	1351.0	1244.2
Criterion 3.6.3	1090.9	1848.8	1766.9
Criterion 3.6.4	825.4	1334.4	1220.0
Criterion 3.6.5	894.8	1498.7	1422.7
Criterion 3.6.6	696.3	1046.6	987.3
Finite element analysis results (adhesive element)			
Specimen	A2	B2	C2
Criterion 3.6.1	1011.8	1642.5	1490.0
Criterion 3.6.2	710.5	1018.7	871.6
Criterion 3.6.3	1002.3	1627.9	1471.5
Criterion 3.6.4	701.1	990.6	857.4
Criterion 3.6.5	815.3	1317.3	1142.9
Criterion 3.6.6	611.3	817.6	765.7

The numerical results given by the one dimensional analysis reveal that the failure loads predicted using linear adhesive properties are much smaller than the measured, while the failure loads predicted using nonlinear adhesive properties are generally close to the measured results. In the case of finite element prediction, the results predicted using the stresses in the center of the adhesive layer element are in a reasonably good correlation with the experimental results. Criterion 3.6.6 appears to give the lower bound of the measured failure loads, and criterion 3.6.1 tends to define the upper bound of the measured failure loads. For the simplified one dimensional analysis, criterion 3.6.2 seems to yield the closest average failure loads, whereas for the finite element analysis, criterion 3.6.5 give the best failure load predictions. It is clear that the three new criteria, i.e. criterion 3.6.2, 3.6.4, and 3.6.6, can provide a good prediction to the failure loads only when they are used in the simplified one dimensional analysis

(b) Composite panel-flange joints

Tong, Sheppard, Kelly et al (1998) investigated adhesive-bonded composite panel-flange joint in compression. Figure 3.21 depicts the typical joint configuration representing a section of stiffened plate and the test set-up (Tong et al 1994). The material used for the specimen is T300/934 graphite/epoxy unidirectional tape with a nominal ply thickness of 0.188 mm. FM300-K film adhesive is used to secondary bond the panel and flanges. The adhesive properties are E=2.4 GPa and ν=0.3. The nominal adhesive thickness is 0.12 mm. Table 3.5 tabulates the lay-ups of the panel and flange for each specimen group. The material properties for each ply are given in Table 3.6. Detailed information on manufacturing and testing results can be found in Tong, Sheppard, Kelly et al (1998).

Table 3.5 Lay-ups of panel and flange in panel-flange joints

Specimen group	Panel	Flange
A	$(\pm45,0,90)_s$	$(\pm45,0,90)_{2s}$
B	$(\pm45,0,90)_s$	$(\pm45,0,90)_s$
C	$(\pm45,0,90)_s$	$(\pm45)_s$
D	$(\pm45,0,90)_s$	$(\pm45)_{2s}$
E	$(0,90,\pm45)_s$	$(0,90,\pm45)_{2s}$
F	$(0,90,\pm45)_s$	$(0,90,\pm45)_s$
G	$(0,90,\pm45)_s$	$(0,90)_s$

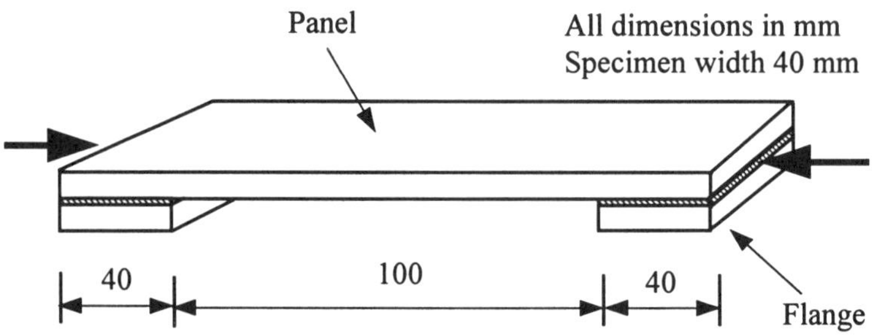

Figure 3.21a Adhesive-bonded skin-stiffener joint in compression

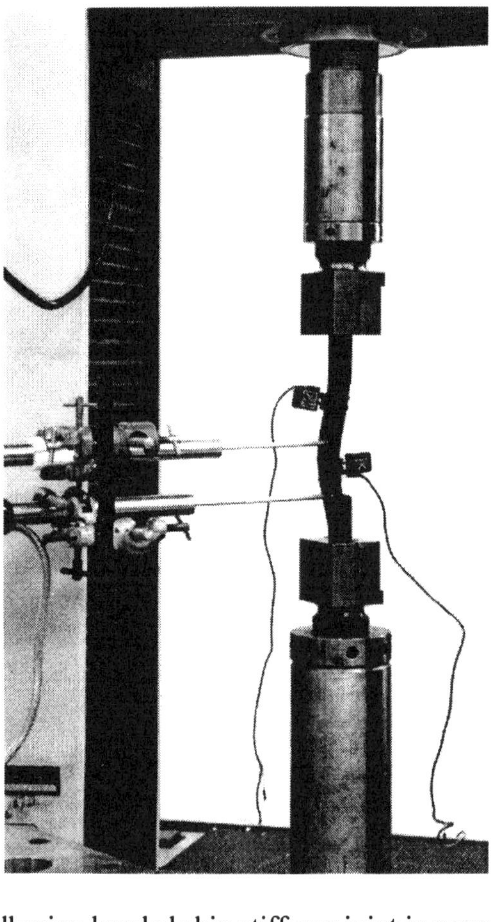

Figure 3.21b Adhesive-bonded skin-stiffener joint in compression

Table 3.6 T300/934 graphite/epoxy unidirectional tape

Mechanical properties	Material strengths
E_L = 127.5 GPa	X_T = 1586 MPa
E_T = 9.0 GPa	X_C = 1517 MPa
E_z= 4.8 GPa	Y_T = 80 MPa
$\nu_{LT} = \nu_{Lz}$ = 0.28	Y_C = 80 MPa
ν_{Tz} = 0.41	Z = 49 MPa
$G_{LT} = G_{Lz}$ = 4.8 GPa	S= 96 MPa
G_{Tz} = 2.55 GPa	R= 34.5 MPa

Finite element analysis is performed for all specimens using MSC/NASTRAN (1992) and the maximum normal stress criterion given in equation (3.4). Because the +45° and -45° plies in the laminates of the bonded joints induce a coupling between bending and twisting, the geometric symmetry of the joints can not be used to reduce the size of the FE models and the assumption of plane strain is invalid. Consequently, a hybrid modeling strategy of Figure 2.22(b), comprising the 4-node quadrilateral shell elements and 8-node solid elements, needs to be used.

Figure 3.22 shows that there is a good correlation between the predicted and measured load-displacement behavior for three specimens from group B. Table 3.7 tabulates the average measured and predicted failure load and displacement for all specimen groups and shows there is a good correlation between the predicted and the average measured failure loads. However, a poor correlation is noted between the predicted and the measured failure axial displacements.

Table 3.7 Summary of the average buckling and failure loads and average deflections (predicted values are given in brackets)

Specimen Group	Buckling load (N)	Failure load (N)	Axial deflection (mm)
A	1083	1188 (1145)	2.3 (1.5)
B	928	1053 (1006)	3.3 (2.5)
C	632	686 (712)	15.7 (7.5)
D	821	923 (921)	4.2 (3.2)
E	1682	2051 (1953)	2.0 (1.2)
F	1469	1672 (1643)	3.0 (1.7)
G	1022	1278 (1310)	9.4 (3.7)

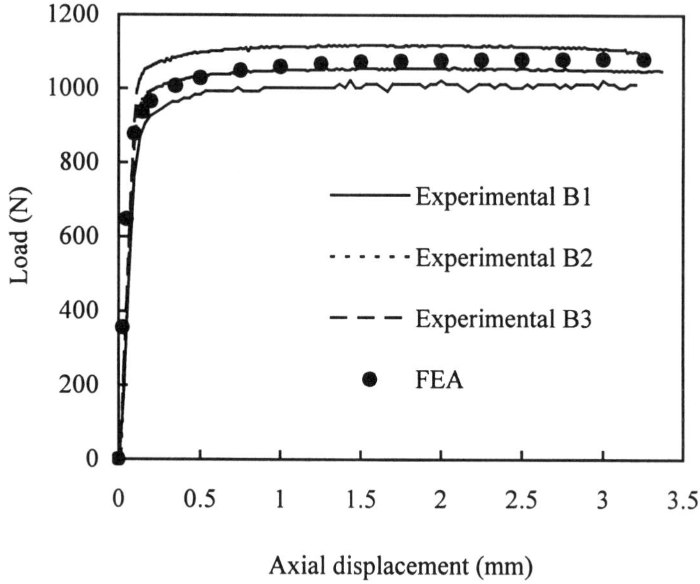

Figure 3.22 Experimental and numerical comparison of load versus axial crosshead displacement for specimen group B (Tong, Sheppard, Kelly et al, 1998)

3.4.2 Examples using zone-based criteria

(a) Aluminum joints (Sheppard, Kelly and Tong, 1998)

Three groups of adhesive-bonded single lap joints were fabricated using aluminum adherends and FM-300 film adhesive. All joints have the same free length of 38 mm, overlap length of 13 mm, width of 25 mm, and adhesive thickness of 0.15 mm, except for the adherend thickness as given in Table 3.8. The adhesive fillets formed at the ends of the joints were removed using a soldering iron. The joints are selected to force a cohesive failure for the purpose of validation of the damage zone model. All joints were tested in an Instron test machine under displacement control. Table 3.8 lists the average failure load for the three joint groups.

In the finite element analysis, a geometric nonlinear analysis procedure was used to correctly capture the large out-of-plane deflections experienced in some of the joints. Material nonlinearity was also incorporated in each analysis to model the stress/strain relation of the adhesive and the aluminum adherends. The Von Mises yield criterion was used for the adhesive layer and the aluminum adherends. It was also assumed that damage occurs when the Von Mises strain attains the ultimate value of 0.2.

Table 3.8 Thickness (mm), average measured and predicted failure load (kN) of adhesive-bonded aluminum single lap joints

Joint group	Adherend thickness	Average measured failure load	Predicted failure load
SLJ1	1.6	10.4	10.4
SLJ2	3.2	11.4	11.9
SLJ3	10.0	13.7	13.4

Three- and four-node plate elements were used to create plane strain models for the joints. Nine elements were used to model the adhesive layer in through-the-thickness direction at the end of the joint. A single 3-noded element is used at the ends of the joints to model the small adhesive fillets and to eliminate the sharp corners at these locations.

Numerical results show that the damage zone size is 0.01 mm^2 corresponding to the average measured failure load of SLJ1 joints. By setting the critical damage zone size to be 0.01mm^2, the failure loads for the remaining two groups, SLJ2 and SLJ3, can be determined. As shown in Table 3.8, the predicted failure loads are within 4% of the average measured failure loads for all joint configurations. This represents a very good correlation with the measured failure loads given the wide range of adherend thickness used for the specimens.

(b) Composite joints (Sheppard, Kelly and Tong, 1998)

Here the six groups of adhesive-bonded composite double lap joints described in Section 3.4.1 are examined again using the damage zone model. In addition, three groups of adhesive bonded single lap joints are fabricated and then tested on an Instron machine under the displacement control. Figure 3.23 shows the dimensions of the single lap joints. Table 3.9 tabulates the lay-ups of the adherends, the overlap length and the adherend thickness for the three groups of single lap joints. The materials used are the same as those used for the double lap joints as given in Section 3.4.1. Table 3.10 summarizes the average measured failure loads for both single lap and double lap joint specimens.

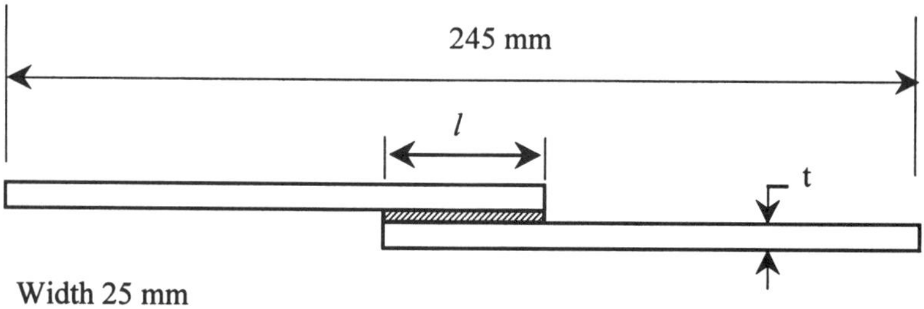

Figure 3.23 Adhesive-bonded composite single lap joint configuration

Table 3.9 Parameters of adhesive-bonded composite single lap joints

Joint group	Adherend lay-up	Overlap length L (mm)	Adherend thickness t (mm)
A1	$(0,90,90,0)_s$	89.4	1.12
B1	$(0_2,90_2,0_2)_s$	134.2	1.68
C1	$(0_2,90_2,90_2,0_2)_s$	178.8	2.24

Table 3.10 Comparison of the average measured and predicted failure loads for composite single- and double lap joints using damage zone model

Joint group	Average measured failure load (kN)	Predicted failure load (kN)
A1	16.25	13.1
B1	21.0	23.5
C1	26.75	23.7
A2/A3	19.85/21.96	19.0
B2/B3	35.60/35.40	35.2
C2/C3	32.11/34.47	35.4

To use the damage zone model for the composite joints, nine through-the-thickness elements were used to model the surface adherend ply and three through-the-thickness elements were used to model the adhesive layer at the end of the joint. In addition, three elements were used to model small adhesive fillets at the ends of the joints. Detailed solution procedure can be found in Sheppard, Kelly and Tong (1998). By assuming that damage occurs at an ultimate strain of 0.01, the critical size of the damage zone was determined to be 0.0018 mm^2. Using this critical damage zone size, failure loads for all specimens can be calculated and are tabulated in the third column in Table 3.10. Results in Table 3.10 shows that there exists a good correlation between the predicted and the average measured failure loads, particularly for joints with thick adherend. For joints with thin adherends, a maximum of 19% difference is noted.

(c) Results using ultimate tensile stress over a zone

Clark and McGregor (1993) proposed a failure criterion referred to as Ultimate Tensile Stress over a Zone. They considered three joint configurations, i.e. single lap, double lap and T-peel joints. The bulk adhesive properties used are E=3076MPa, ν=0.3 and the Ultimate Tensile Stress = 70MPa. For the adhesive fillet end shown in Figure 3.24, it was postulated that failure occurs when zone size of the maximum principal stress exceeding 70 MPa attains 0.68mm. The critical zone size was determined using the single lap joints, and then used to predict the failure loads for other joints made of the same adhesive and adherend materials. Table 3.11 compares the predicted and measured failure loads for the double-strap and T-peel joints. A good correlation is found for different joint configurations.

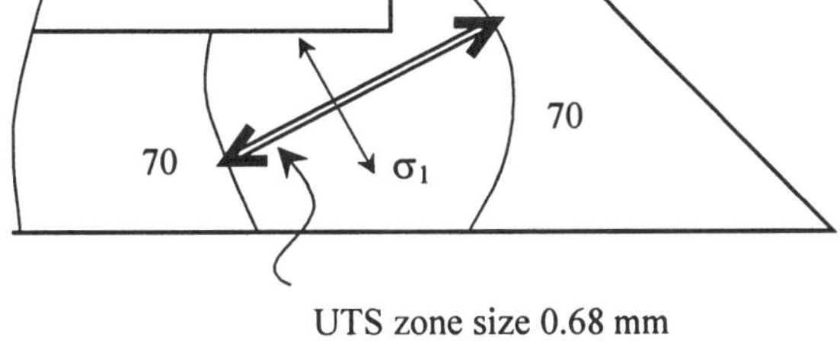

Figure 3.24 Schematic of UTS zone size of 0.68mm.

Table 3.11 Comparison of predicted and measured strengths (N/mm) of various joint configurations (Clark & McGregor, 1993)

Joint configuration	Predicted strength	Measured strength
Single lap	Calibration	320
Double-strap	701	740
T-peel	123.7	123

3.5 APPLICATIONS OF FRACTURE MECHANICS CRITERIA

For the balanced single lap and lap shear joint under tension as shown in Figure 3.25, we can determine the longitudinal membrane forces and the bending moments in the continuous adherends at the overlap ends.

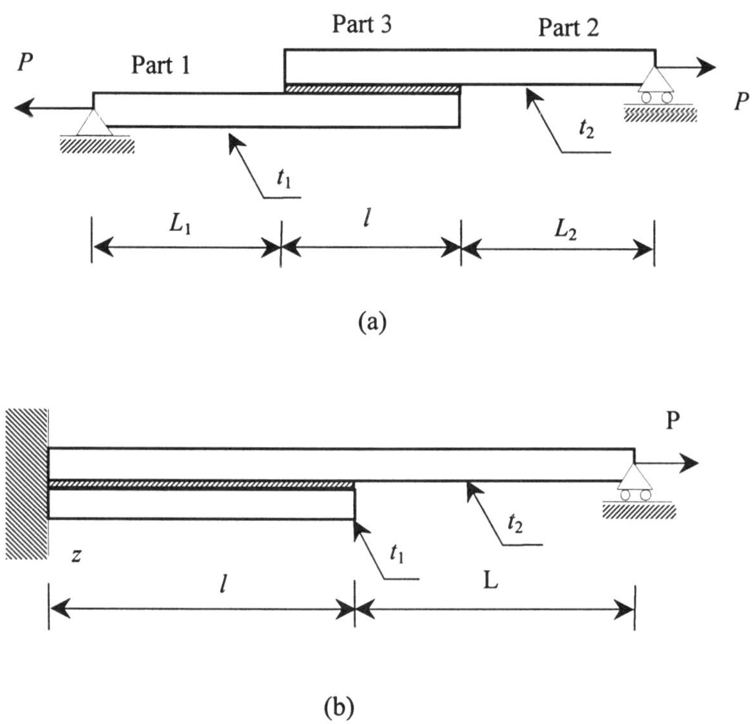

Figure 3.25 Configuration for a single lap and a lap shear joint under tension loading (a) Single lap joint, and (b) lap shear joint

The energy release rates of mode I and II at both ends can be calculated using the longitudinal force and bending moment at the overlap end and the following equations:

$$G_{II} = \frac{A_1}{16}\left(\frac{P}{A_1} + \frac{t_1 M_i}{2D_1}\right)^2, \qquad G_I = \frac{M_i^2}{4D_1} \tag{3.41}$$

where A_1 and D_1 are the longitudinal membrane stiffness and bending stiffness of each adherend. M_i (i=1,2) is the bending moments at the overlap ends overlap acting on the top adherend at 1 and 2 respectively (Tong, 1996). By using the failure criteria given by Equation (3.22) and (3.23) and the measured in-situ envelopes (Fernlund et al 1994 and Papini et al 1994), joint strength can be determined following an iterative numerical procedure (Tong, 1998a).

Table 3.12 Comparison of predicted and measured fracture loads (N/mm) for equal adherend lap shear joints with t_1=t_2= 12.54 mm and a width of 20 mm

Geometry (mm)		Measured and predicted fracture load using various types of criteria (see equations 3.22-23)			
L	*l*	Test*	In-situ**	Linear	Quadratic
160	187	1100	971 (-11.7)	965 (-12.3)	1174 (6.7)
165	185	968	962 (-0.6)	955 (-1.3)	1164 (20.3)
187	159	876	883 (0.8)	876 (0.0)	1071 (22.3)
182	154	915	874 (-4.5)	866 (-5.4)	1059 (15.7)
162	180	935	953 (1.9)	947 (1.3)	1153 (23.3)
143	197	1092	1013 (-7.2)	1008 (-7.7)	1221 (11.8)
213	127	726	786 (8.3)	777 (7.0)	951 (31.0)
154	192	1044	988 (-5.4)	982 (-5.9)	1194 (14.4)
143	196	1081	1011 (-6.5)	1006 (-6.9)	1218 (12.7)
220	118	769	758 (-1.4)	750 (-2.5)	916 (19.1)
133	210	1226	1055 (-13.9)	1050 (-14.4)	1267 (3.3)
187	157	966	878 (-9.1)	871 (-9.8)	1065 (10.2)
242	101	720	703 (-2.4)	695 (-3.5)	846 (17.5)
168	184	960	957 (-0.3)	950 (-1.0)	1158 (20.6)
240	112	695	735 (5.8)	727 (4.6)	888 (27.8)

* Data taken from Fernlund et al (1994)

** Results obtained using failure envelopes given by Fernlund et al (1994)

Table 3.12 gives the fracture loads for the lap shear joints predicted using the present two failure criteria and compares with the measured fracture loads taken from Fernlund et al (1994). Between the two criteria, the linear criterion tends to give the smallest average difference 5.6%, while the quadratic criterion yields an average difference of 17.1%. Thus the linear criterion is better for predicting the strength for the lap shear joints. This is because the linear criterion fits best to the measured failure envelope of Figure 3.13 when the phase angle varies from 49.1° to 90° for the lap shear joints (Papini et al, 1994).

Table 3.13 Comparison of predicted and measured fracture loads (N/mm) for equal adherend single lap joints (t_1=t_2= 12.5 mm, width was 20 mm).

Geometry (mm)			Measured and predicted fracture load using various types of criteria (see equation 3.22-23)			
L_1	l	L_2	Test*	In-situ**	Linear	Quadratic
141	86	125	572	553 (-3.3)	546 (-4.5)	646 (12.9)
144	84	125	555	545 (-1.8)	539 (-2.9)	638 (14.9)
178	82	141	547	522 (-4.6)	515 (-5.8)	610 (11.5)
176	86	138	525	527 (0.4)	521 (-0.8)	618 (17.7)
245	75	80	438	419 (-4.3)	414 (-5.5)	492 (12.3)
265	55	80	391	385 (-1.5)	380 (-2.8)	449 (14.8)
216	98	88	497	475 (-4.4)	469 (-5.6)	561 (12.9)
203	62	135	551	470 (-14.7)	464 (-15.8)	549 (-0.4)
185	109	97	544	521 (-4.2)	514 (-5.5)	615 (13.0)
219	74	96	469	445 (-5.1)	439 (-6.4)	522 (11.3)
167	127	96	565	565 (0.0)	558 (-1.2)	670 (18.6)
134	56	108	530	491 (-7.4)	485 (-8.5)	570 (7.5)
202	105	57	496	453 (-8.7)	447 (-9.9)	533 (7.5)
196	111	94	559	514 (-8.1)	507 (-9.3)	607 (8.6)
216	92	82	525	458 (-12.8)	452 (-13.9)	540 (2.9)
148	115	106	574	576 (0.3)	569 (-0.9)	679 (18.3)
148	115	106	595	576 (-3.2)	569 (-4.4)	679 (14.1)
148	115	106	587	576 (-1.9)	569 (-3.1)	679 (15.7)

* Data taken from Fernlund et al (1994)

** Results obtained using failure envelopes given by Fernlund et al (1994)

Table 3.13 compares the fracture loads predicted using the present two failure criteria and the measured fracture loads for the single lap joints made of the 7075-T6 aluminum adherends bonded with Cybond 4523GB epoxy adhesive system (Fernlund et al, 1994). It is noted that the quadratic criterion gives the average difference of 11.9%, and the linear yields an average difference of 5.9%. It is thus found that the linear criterion seems to be the appropriate one for these joints.

In Table 3.14, the measured and predicted fracture loads are tabulated for the single lap joints made of the 7075-T6 aluminum adherends bonded with Permabond ESP 310 adhesive system (Papini et al, 1994). It is shown that the linear criterion gives an average difference of 13.4%, while the quadratic results in an average differences of 5.4% for these joints. Clearly the quadratic criterion seems to be appropriate for predicting the strengths of the joints manufactured using this adhesive system.

Table 3.14 Comparison of measured and predicted fracture loads (N/mm) for equal-adherend single lap joints (t_1=t_2=12.7 mm).

Geometry (mm)			Measured and predicted fracture loads using various types of criteria (see equations 3.22-23)			
L_1	l	L_2	Test*	In-situ**	Linear	Quadratic
142	80	140	1408	1514 (7.5)	1289 (-8.5)	1429 (1.5)
142	80	140	1352	1514 (12.0)	1289 (-4.7)	1429 (5.7)
142	80	140	1361	1514 (11.2)	1289 (-5.3)	1429 (5.0)
142	80	140	1405	1514 (7.8)	1289 (-8.3)	1429 (1.7)
142	80	140	1450	1514 (4.4)	1289 (-11.1)	1429 (-1.4)
227	81	148	1478	1452 (-1.8)	1222 (-17.3)	1361 (-7.9)
227	81	148	1481	1452 (-2.0)	1222 (-17.5)	1361 (-8.1)
227	81	148	1524	1452 (-4.7)	1222 (-19.8)	1361 (-10.7)
254	54	148	1379	1284 (-6.9)	1083 (-21.5)	1195 (-13.3)
249	59	148	1373	1315 (-4.2)	1108 (-19.3)	1225 (-10.8)
227	64	171	1284	1367 (6.5)	1157 (-9.9)	1279 (-0.4)
172	117	172	1654	1721 (4.1)	1458 (-11.9)	1638 (-1.0)
200	90	171	1477	1534 (3.9)	1297 (-12.2)	1445 (-2.2)
237	53	171	1308	1301 (-0.5)	1102 (-15.7)	1214 (-7.2)
197	82	192	1493	1496 (0.2)	1268 (-15.1)	1408 (-5.7)
166	140	165	1768	1867 (5.6)	1581 (-10.6)	1791 (1.3)
203	104	164	1647	1615 (-1.9)	1362 (-17.3)	1526 (-7.3)
230	80	164	1457	1457 (0.0)	1229 (-15.6)	1367 (-6.2)

* Data taken from Papini et al (1994)

** Results obtained using failure envelopes given by Papini et al (1994)

3.6 FATIGUE STRENGTH PREDICTION

Any rationally designed structure must possess an adequate service life under the operating conditions to be experienced by the structure. When subjected to fluctuating loads, materials or structures can fail or fracture at stress or strain levels much lower than they can withstand under monotonic or static loading. This type of failure or fracture phenomenon of a material or structure subjected to oscillatory dynamic loading is often referred to as fatigue. Factors affecting structural fatigue behavior include the range and frequency of the applied load, materials, geometrical parameters, and the operating environments, for example, exposed to elevated temperature, moisture etc.

Fatigue is definitely an important issue to be addressed in any rational design of bonded joints or structures. To be able to design a bonded joint to a specific service life, it is of prime importance to (a) understand the mechanisms of fatigue failure, (b) develop reliable test methods that evaluate adhesive systems and quality assurance processes, and (c) predict quantitatively the expected service life. Fatigue testing of bonded joints can be very expensive especially when various loading conditions are included, such as the range and frequency of loading. Evaluation of fatigue life for a bonded joint or structure becomes even more difficult and demanding. Firstly, the structural performance of a bonded joint may be adversely affected when exposed to aqueous environments (Kinloch, 1987), especially a hot-wet environment (i.e. at an elevated temperature and moisture or immersed in water), and these extreme environment may be frequently encountered by a bonded structure under repeated loading. Secondly, quantitative stress analysis of bonded joints represent one of the most challenging difficult tasks in structural design due to the presence of bimaterial interfaces, cracks, geometrical uncertainties and nonlinear material properties.

Many researchers have investigated the fatigue behavior of adhesives and bonded joints using fracture mechanics methods (Kinloch, 1987). Prediction of the fatigue life of a bonded joint or component may be classified into two major aspects. The first aspect is to determine the crack growth rate function of an adhesive system in terms of the strain energy release rate under given loading conditions. The second aspect is to develop a model that establishes the relationship between the strain energy release rate and the crack length for a bonded joint or component, and then to integrate the energy release rate function with respect to the crack length from its initial value to the value at which catastrophic fast crack propagation starts.

3.6.1 Determination of crack growth rate

In order to predict the fatigue life of a bonded joint or component, it is necessary to obtain a relationship between the crack growth rate and the corresponding energy release rate. A general crack growth rate function can be written as

$$\frac{da}{dN} = C(\Delta G) \tag{3.42}$$

where $C(\Delta G)$ is a function of the energy release rate range ΔG, which represents the range of the applied strain energy release rate ΔG given by

$$\Delta G = G_{max} - G_{min} \tag{3.43}$$

where G_{max} and G_{min} are the maximum and minimum values of the strain energy release rate produced due to the applied loading per cycle (Paris and Erdogan, 1960; Paris et al, 1961).

A modified form of the crack growth rate function in equation (3.42) is given as

$$\frac{da}{dN} = C(G_{max}) \tag{3.44}$$

where the maximum value of the strain energy release rate G_{max} is used rather than the range of the energy release rate ΔG.

Mostovoy and Ripling (1975) employed a linear-elastic fracture mechanics (LEFM) approach and successfully described the fatigue crack growth behavior of a range of epoxy-based adhesives sandwiched between aluminum-alloy substrates. In their tests, a tapered double cantilever-beam (TDCB) joint specimen shown in Figure 3.11 was used. The width of the specimen is b, The crack length is a, and the height of the arms at the crack tip is d. The two tapered arms are loaded with a pair of loads P in opposite directions to produce crack opening or mode I crack propagation at the sharp precrack tip along the bond line. The experimental results given by

Ripling et al (1971, 1983, 1988) and Mostovoy and Ripling (1975) show that the mode I crack growth rate, which was defined as the rate of crack growth per cycle, da/dN, may be expressed in terms of the range of the strain energy release rate ΔG due to the applied loading as follows:

$$\frac{da}{dN} = A(\Delta G)^n \tag{3.45}$$

where A and n are material constants. They found that ΔG has some limiting threshold value G_{th}. When ΔG approaches G_{th} the crack propagates very slowly or almost stops growing. On the other hand, the crack propagated very fast as ΔG approaches to the typical value of the critical strain energy release rate, G_c, for crack growth under short-term monotonic loading conditions.

Ripling et al (1988) conducted testing for FM 300K.05 and FM 73.06 adhesives sandwiched between 7075-T651 aluminum alloy using the TDCB specimen as schematically shown in Figure 3.11 at different cyclic loading frequency and relative humidity. Table 3.15 tabulates the two constants at two cyclic loading frequencies and the dry and wet environments. The results of the two constants A and n show clearly that the cyclic loading frequency and the relative humidity have an important effect on the crack growth rates.

Table 3.15 Crack growth rates of FM 300K .05 and FM 73.06 adhesives measured with TDCB specimens (after Ripling et al, 1988)

Adhesive	Frequency (Hz)	Relative humidity (%)	A	n
FM 300K.05	10	40	$1.32\text{x}10^{-14}$	4.35
FM 300K.05	3	80-90	$3.90\text{x}10^{-17}$	5.33
FM 73.06	10	30	$4.20\text{x}10^{-14}$	3.99
FM 73.06	3	75	$4.15\text{x}10^{-17}$	5.13

A more complicated expression for expressing the crack growth rate in terms of the induced strain energy release rate was proposed by Martin and Murri, (1990). Their formula has five material constants to be determined experimentally. The formula is given as:

$$\frac{da}{dN} = AG_{\max}{}^{n} \left\{ \frac{1 - \left(\frac{G_{th}}{G_{\max}} \right)^{n_1}}{1 - \left(\frac{G_{\max}}{G_c} \right)^{n_2}} \right\} \tag{3.46}$$

where G_{th} is the minimum or threshold value of the applied strain energy release rate, below which crack growth is very slow. A, n, n_1 and n_2 are constants for a given set of test conditions, such as frequency. G_C is the critical strain energy release rate under short-term loading conditions.

Curley et al (1998) used the maximum value of the strain energy release rate G_{max} rather than the range of the strain energy release rate ΔG, expressed the crack growth rate as follow:

$$\frac{da}{dN} = AG_{\max}{}^{n} \tag{3.47}$$

where A and n are empirical material constants for a given loading ratio, frequency of tests and environment.

The J-integral has been applied to fatigue accompanied by large scale yielding by assuming a growth law of the form (Dowling and Begley, 1976; Lamba, 1975; Lambert et al, 1988).

$$\frac{da}{dN} = f(\Delta J, R) \tag{3.48}$$

where ΔJ is contour integral for cyclic loading, namely

$$\Delta J = \int_{\Gamma} (\Delta w dy - \Delta T_i \frac{\partial \Delta u_i}{\partial x} ds) \tag{3.49}$$

and

$$\Delta w = \int_0^{\Delta \varepsilon_{KD}} \Delta \sigma_{ij} d(\Delta \varepsilon_{ij}) \tag{3.50}$$

If ΔJ varies during cyclic loading, the crack growth in a given cycle may depend on the loading history as well as the current ΔJ value. The above equation is valid in the case of constant amplitude fatigue in small scale yielding, because of the relationship between J and K under linear elastic conditions. In general, ΔJ cannot be interpreted as the range of the applied J value. In the presence of significant plasticity the validity of the above equation is less clear. Dowling and Begley (1976) proposed the following power law expression in ΔJ

$$\frac{da}{dN} = C\Delta J^m \tag{3.51}$$

for fatigue crack growth under large scale yielding conditions where K is no longer valid.

3.6.2 Prediction of fatigue life

When the relationship between the crack growth rate and the full range energy release rate is established for a given adhesive system under a given loading conditions, the fatigue life can be evaluated via the following two steps:

(a) Develop a model to calculate the energy release rate of a bonded joint in terms of the crack length;

(b) Integrate the energy release rate function with respect to the crack length.

The first step has been discussed in the previous section. Here we focus on the second step. Integrating equation (3.42) with respect to the crack length a from its initial crack length a_0 to its critical value a_f at which catastrophic fast crack propagation occurs, the fatigue life can be determined by

$$N_f = \int_{a_0}^{a_f} \frac{1}{C(G)} da \tag{3.52}$$

where G is a function of the crack length a, i.e. $G(a)$.

As given by Curley et al (1998), the fatigue behavior for the "XD4600" adhesive/aluminum-alloy joints can be described by the following fitted curve.

$$\frac{da}{dN} = 5.86 \times 10^{-18} G_{max}^{\ \ 3.54} \left\{ \frac{1 - \left(\dfrac{355}{G_{max}} \right)^{16.0}}{1 - \left(\dfrac{G_{max}}{3500} \right)^{0.46}} \right\}$$

The XD4600 is a one-part epoxy-paste adhesive (Grade XD4600 supplied by Ciba Polymers, UK, the Aluminum-alloy contains 4.0 to 4.9% magnesium and 0.4 to 1.0% manganese Grade: ENAW -5083). The above crack growth rate is schematically plotted versus the maximum energy release rate in Figure 3.26. It is clear when G_{max} is less than or close to 355 the crack propagates very slowly, and when G_{max} is approaching 3500 the crack growth rate is high and a fast crack growth is then expected.

Curley et al (1998) employed both analytical and finite element analysis methods to establish the relationship between the maximum strain energy release rate and the crack length. In their analytical model, the bending moments in the continuous adherend at the crack tip was used to calculate the mode I strain energy release rate. The bending moment expression given by Hart-Smith was used. It was assumed two cracks or debonds propagate at the same rate from the two overlap ends towards the center of the overlap. Thus effective length $2c$-$2a$ is used to replace $2c$ (c is half overlap length and a is crack length at each overlap end). For a single lap joint, the strain energy release rate is a function of the overlap length and the applied load. For a given overlap length, the monotonic failure load T_c can be determined corresponding to a static fracture toughness G_c. However, if the applied load P is decreased, to adhesive G_c, the overlap length needs to be decreased. The amount of reduction in the overlap length can be defined as the critical crack length. This may be explained as follows; when the cracks propagate towards the overlap center, the static fracture toughness for the cracked single lap joint with a shorter overlap increases and may attain to the static fracture toughness at a critical crack length of the applied load is reasonable high. Obviously, if the applied load is sufficiently low, critical crack length may not be achievable, and thus fatigue failure may not occur at all. The number of cycles to failure in the single lap joints is expressed as a function of the maximum applied load using the analytical model. It is shown that there exists a poor agreement between the predicted and the measured numbers of cycles to failure, namely, the predicted fatigue life is significantly lower than the measured one.

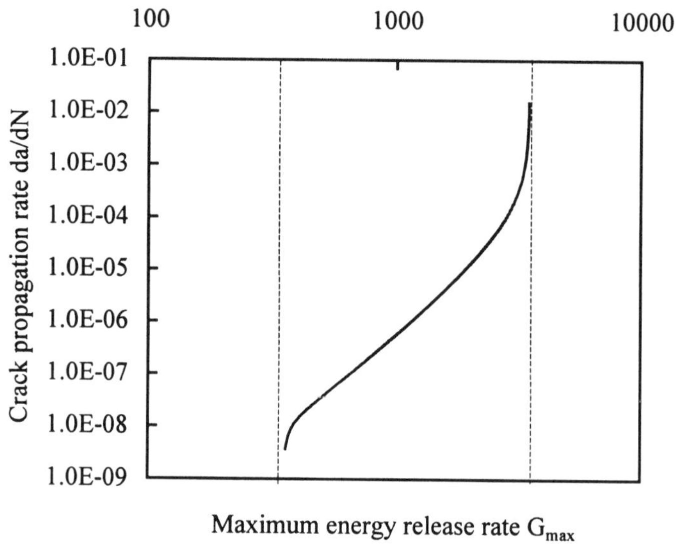

Figure 3.26 Crack growth rate versus the maximum energy release rate

In the part of fatigue life prediction using finite element model, 2-D plane strain analysis is conducted in conjunction with both virtual crack closure method and the J-integral method for evaluating the values of energy release rate. Curley et al (1998) found that a good agreement between the predicted and measured numbers of cycles to failure is observed for the finite element analysis model.

4

Damage tolerance

4.1 INTRODUCTION

Damage tolerance is another important issue in structural design and in addressing structural integrity and durability. Damage tolerance design has been applied in many real engineering structural designs, such as aircraft structural design. In damage tolerance design, some of the key questions raised and answered are does the presence of damage affect the structural integrity? What is the remaining service life? For example, micro cracks may initiate and propagate around rivet holes in aircraft fuselage lap connections. It can be asked, does the presence of micro cracks around rivet holes affect the structural integrity? Can the structure survive till next inspection? What is the service life remained? Fracture mechanics has been used to address this issue (Atluri, 1997).

Advanced composite materials are being used widely in aircraft structures due to its high specific stiffness and strength. It is also well known that laminated composites are prone to impact damage. Low energy impact can produce barely visible impact damage and can reduce the compressive strength by up to 50% comparing to the undamaged. There exists an enormous amount of research publications available in the field of damage tolerance of advanced composite materials. Readers who requires more information on this subject are referred to the book by Springer and Finn (1991), Sierakowski and Newaz (1995) for detailed presentation of damage tolerance assessment and characterisation methods as well as the role of damage tolerance in the design of composite structures.

Adhesive bonding is being increasingly used to join composite member to form bonded structures, such as secondary bonded stiffened plates and shells. Damage tolerance issue exists for such adhesive bonded joints, namely, is the structural bonded joint tolerant to damage in or around joining area? what is the remaining life of the damaged joint? While this chapter is not intended to answer all these important questions fully, we intend to present an overview on the damage tolerance issue of bonded joint and to show its importance.

Damage can occur in the adhesive and/or in the adherends. Voids or defects in adhesive may be developed due to air bubbles trapped in the adhesive or foreign objects left during bonding process. Another possible source of damage is weak bonding or kissing bond or debonding due to improper surface preparation and incorrect material handling and bonding process. Hart-Smith (1978), Kan and Ratwani (1983), and Rossetos and Zhang (1993) have studied the effect of adhesive debonding or void in adhesive on stress distributions in bonded joints.

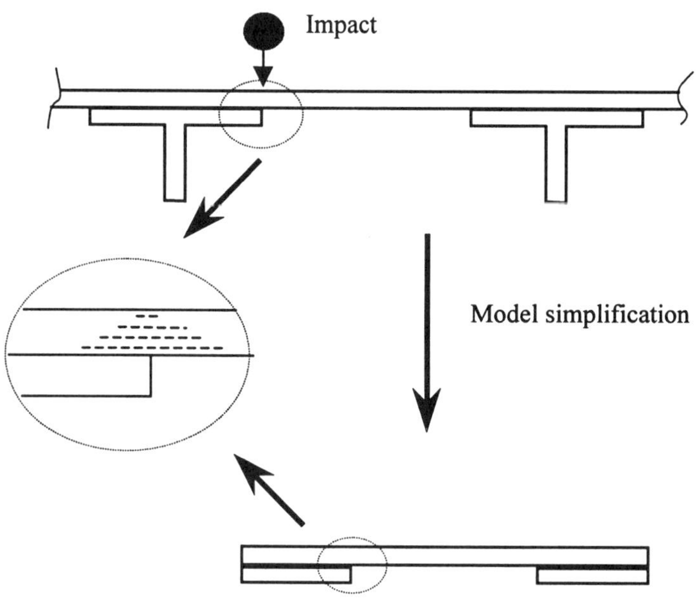

Figure 4.1 Scenario of impact damage to a stiffened composite panel at the overlap end

Damage can also exist in adherends, especially in composite adherends. Figure 4.1 shows a scenario of impact damage occurred to a stiffened composite panel near the flange end of the T stiffener. This scenario shares the same idea as laminated composite panel subjected to impact damage, such as due to dropped tools. Although there has been a lot of studies investigating and understanding damage tolerance of composite materials and failure behaviour of adhesive bonded composite joints as two separate issues, there has been little effort devoted to addressing damage tolerance issue of adhesive bonded joints. Recently, Tong, Zhu and Steven (1998) and Tong (1998b) have demonstrated the effect of embedded cracks on behaviour of single lap joints.

In this chapter, we will first discuss the effect of defects or voids in adhesive on stress distribution in structural bonded joints. Secondly, for bonded composite joints, recent numerical and experimental results will be presented to illustrate the effect of embedded cracks in composite adherends on static strength. Lastly, a brief recommendation on further research topics will be presented. It worth pointing out that damage tolerance study of bonded joint can be regarded as an alternative perspective in understanding failure mechanism and predicting failure strength.

4.2 EFFECT OF DEBOND OR VOIDS IN ADHESIVE

Defects or voids may exist in adhesive due to inappropriate surface preparation and bonding process. For example, air bubbles may be formed by entrapped air in adhesive. Gaps may also exist in large or complex structural bonding due to inadequate quality control of adhesive application. For example, end gaps may exist in an adhesive bonded stepped lap joint. Figure 4.2 shows a cross-sectional view of some typical defects in adhesive in bonded structures. As shown in the figure, a gap refers to an area where adhesive is not applied for some reasons while a debonding refers to an interfacial crack unzipped due to either local weak bonding or local stress concentration.

Hart-Smith (1978) presented an analysis on the adhesive stresses and strains at discontinuities and cracks in bonded structures. Based on an elastic-plastic shear stress analysis of bonded discontinuous members as shown in Figure 4.3, Hart-Smith investigated the effect of debonding on adhesive stresses and strains for three bonded structural configurations. These are: (a) a sheet bonded to cracked stiffener, (b) central stiffener bonded to fully cracked sheet, and (c) central stiffener bonded to a partly cracked sheet with equal crack length on each side of the stiffener. Hart-Smith's analysis focuses on conditions of debonding development between stiffener and sheet rather than on conditions of stable and unstable crack growth (Atluri, 1997). For typical aerospace structures, debonding can occur followed by self-arrest and will not cause catastrophic unzipping.

However, for the second case, central stiffener bonded to a fully cracked sheet, unstable unzipping between stiffener and sheet can take place. Hart-Smith also suggested that stiffener run-out be tapered to reduce stress concentrations based on his analysis.

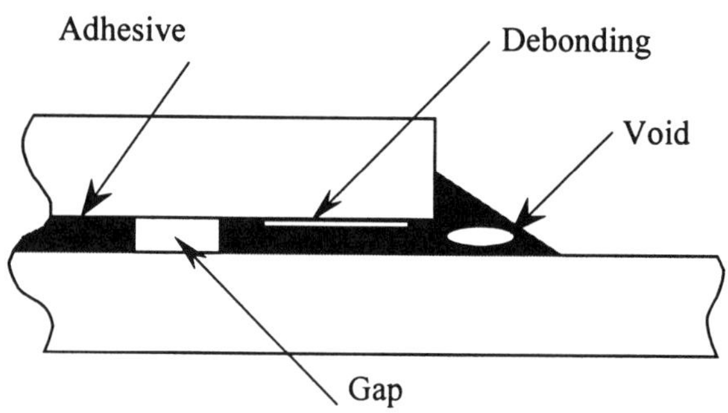

Figure 4.2 Typical defects in adhesive in bond structures

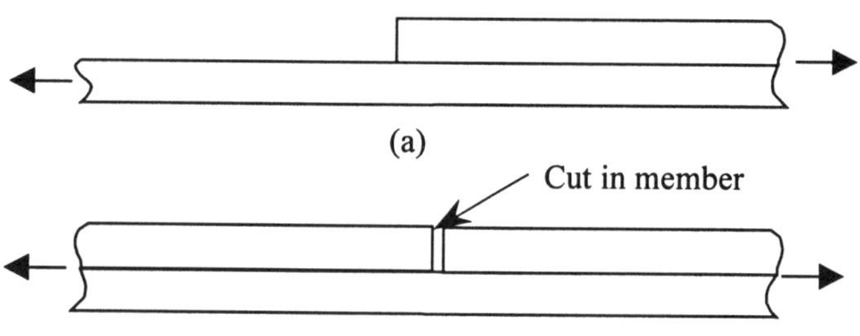

Figure 4.3 cracked skin induced debonding in adhesive

Kan and Ratwani (1983) investigated the influence of end gap between two members and flaws in adhesive on the stresses in metal-to-composite stepped lap joint. Figure 4.4 depicts the two types of defects, debonds in the adhesive and gaps between step ends. In fabrication process, the gaps between step ends are normally filled with adhesive and then axial force can be transferred between two step ends. However, due to the composite layer not butting against the metallic part, end gaps sometimes occur. Based on shear lag theory, they found that the presence of step-end gaps and debonding flaws affects the stress distributions in both adherends and the adhesive in the vicinities of the flaws.

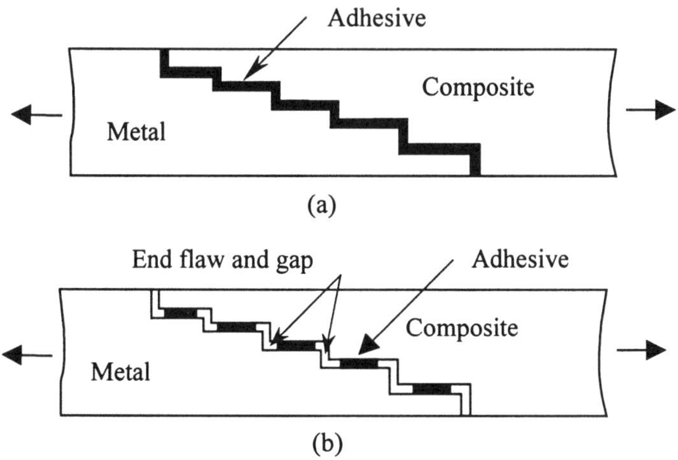

Figure 4.4 Metal-to-composite adhesive bonded stepped lap joint. (a) perfectly bonded joint; (b) joint with gaps and flaws near each step end.

Rossetos and Zang (1993) studied the effect of adhesive void in the overlap on the stress distribution in a bonded joint as shown in Figure 4.5. Similarly to Kan and Ratwani (1983), their analysis only considers the shear stresses with the peel stress being neglected. They found that a void located in the centre of joint does affect the peak stress. However, when the void is located close to one end where a high stress gradient already exist, it causes noticeable increase in peak stress.

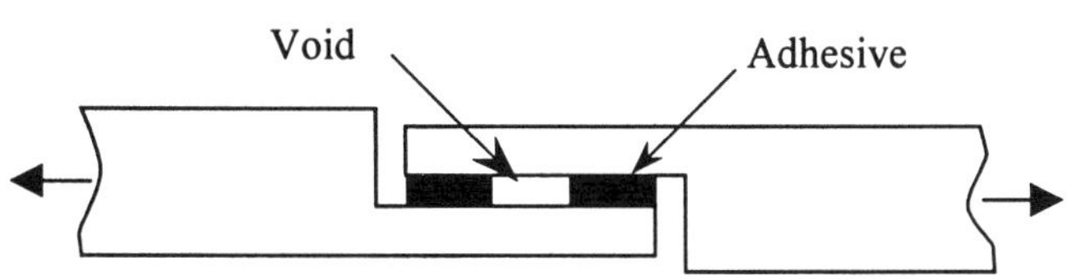

Figure 4.5 gaps or voids in the overlap or at the end of the overlap in lap joints

4.3 EFFECT OF DAMAGE IN ADHERENDS

Damage tolerance is a major issue in designing advanced composite structures. To address the damage tolerance issue when damage exists in composite adherends, let us consider the scenario of stiffened plates presented in Figure 4.1. Stiffened composite panels and shells can be made by secondary bonding the stiffeners onto the thin skin plates and shells, which represent typical aircraft structures, for example, aileron, rudder and wing surfaces. Such types of composite structures are prone to impact damage. When a dropped tool or hailstones hit the plate, delamination damage can occur, as schematically depicted in Figure 4.1, delaminations can occur between several plies. This is a three dimensional damage tolerance issue. To simplify the problem, let us consider the simplified section of the stiffened plates, i.e. a plate with a doubler (adherend) at each end, as shown in Figure 4.1. Tong, Sheppard and Kelly (1994) and Tong, Sheppard, Kelly and Chalkley (1998) have studied such type of bonded structure without including the delamination damage when it is subject to compressive load. Half of the simplified structure in Figure 4.1 is the typical lap shear joint (LSJ) configuration. Thus a lap shear joint configuration with an embedded crack is chosen to illustrate the effect of delamination damage. For the purpose of comparison, a single lap joint (SLJ) with two embedded cracks is also selected. The two joint configurations are shown in Figure 4.6. In the lap shear joint (LSJ), there is one embedded crack in the continuous adherend, while in the single-lap joint there are two cracks placed in the continuous adherends at both ends of the overlap.

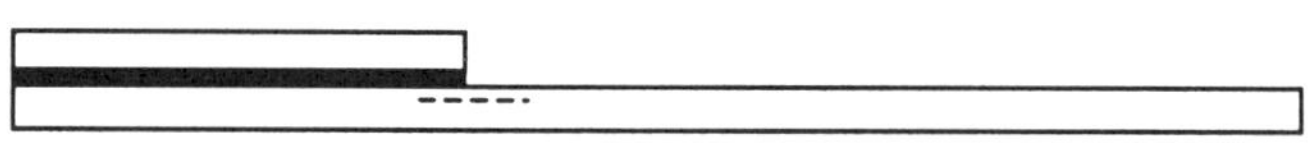

(a)

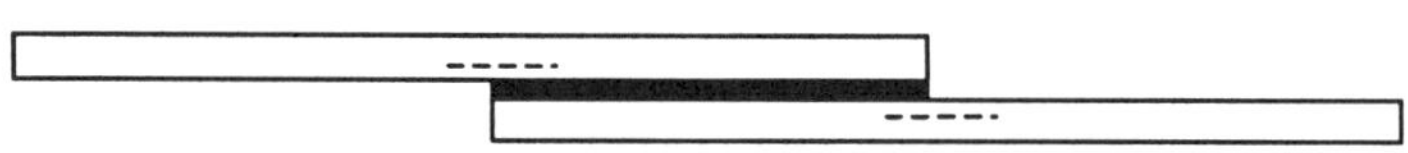

(b)

Figure 4.6 Lap shear and single lap joints with embedded cracks indicated by dashed lines.

Another simplification is to consider only one delamination between the surface and second ply in the continuous adherend instead of placing several delamination cracks between different plies. In doing so, we can investigate the effect of the embedded cracks on the failure mechanism and failure strength of lap joints, and thus address the damage tolerance issue. It is well know that interlaminar delamination is a typical failure mode observed in an adhesive-bonded composite lap joint (Adams, 1989, Kairouz and Matthews, 1993). The two major contributing factors are considered to be the high gradient of peel stress at overlap ends and the interlaminar weakness of composite adherends. Investigation of embedded crack effects can also provide us with an alternative method for predicting joint strength.

4.3.1 Specimen and testing procedure

Figure 4.7 shows the major dimensions of the single lap and lap shear joint specimens. All single lap and lap shear specimens have the same width of 25.4mm, the same adherend thickness of 1.72 mm and the same bondline thickness of 0.16 mm. For all specimens, the cracks are embedded between the surface and the second ply 0.215 mm beneath the surface. For the single lap joints, the crack length and the relative location of the two cracks are varied to form six specimen groups as tabulated in Table 4.1. There are 5 specimens in each group except that group SLJ-0 and SLJ-6-40 have only 3 and 4 specimens respectively. The first four specimen groups are designed to investigate the effect of the crack length while the last two are selected to study the effect of relative location, within and outside the overlap zone. For the lap shear joint specimens, only the crack length is varied from 0 to 10 mm. There are five specimen groups, LSJ-0, LSJ-2, LSJ-4, LSJ-8 and LSJ-10, in which the last digit indicates the crack length. For example, LSJ-4 represents a specimen with an embedded crack of 4 mm long.

Table 4.1 Location and size of cracks embedded in single lap joints

Specimen group	Offset distance d (mm)	Crack length a (mm)
SLJ-0	50	0
SLJ-4	50	4
SLJ-6	50	6
SLJ-8	50	8
SLJ-6-60	60	6
SLJ-6-40	40	6

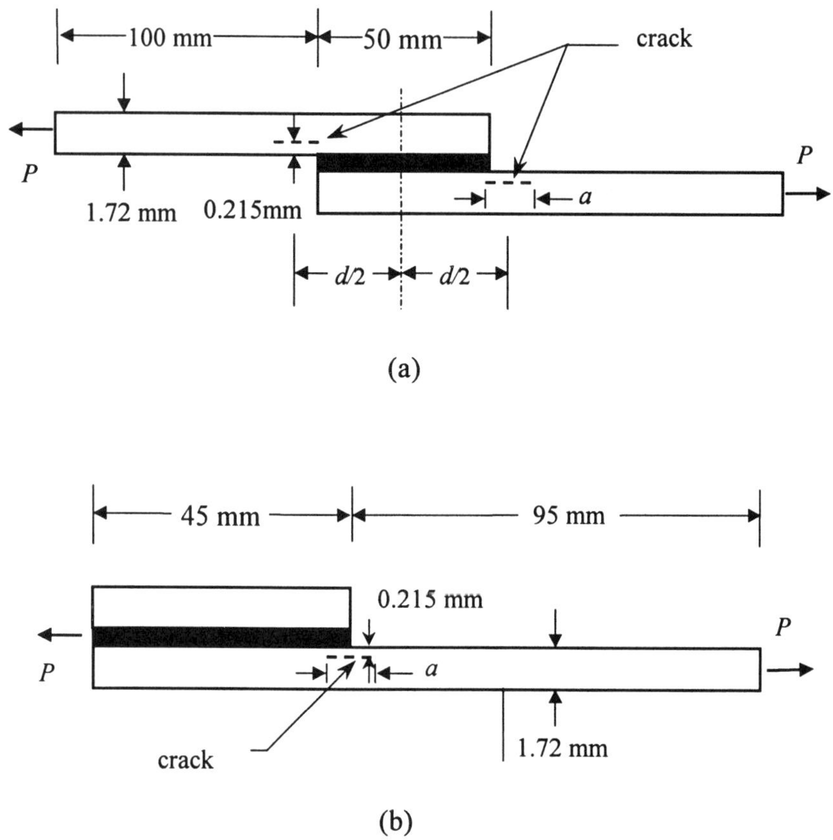

Figure 4.7 Dimensions of the lap shear and single lap joint specimens

As described in detail in Tong (1998b), the material used is T300/934 carbon/epoxy plain woven prepreg, which has a curing temperature of 177°C and possesses the same mechanical properties in both warp and weft directions. Composite panels were manufactured by laminating 8 plies of the plain-woven prepreg in 0° direction, i.e. $(0)_{8S}$. Each ply had a nominal thickness of 0.215 mm. All panels were then cured in an autoclave following manufacturer's recommended curing cycle (Anonymous, 1989).

In the lay-up process, two plies of 0.02 mm thick Teflon film of different width were embedded between the surface and second ply to model cracks of different length at the chosen locations. It is worth pointing out that although the use of Teflon film to model cracks is a common practice, it can be argued that such a practice tends to form a notch rather than a sharp crack.

In the curing process, care must be taken to ensure that all the Teflon films stay at the right position. FM300-K film adhesive was then used to bond the two panels following a moderate hand sanding with sandpaper and an acetone wash. Metal shims were used to achieve a uniform bondline thickness of 0.16 mm. Specimens were prepared by cutting the panels in such a way that there were no metal shims left in the bonding area. Aluminum end taps of size 25.4 mm by 40 mm were used for both types of joints to overcome the eccentricity introduced when installing a specimen on testing machine.

All specimens were loaded in axial tension on an Instron testing machine at a loading rate of 0.5 mm/min. The applied load and the cross-head displacement were recorded.

4.3.2 Results and discussion

(a) Failure loads

For the single lap joint specimens, the measured axial load increases almost linearly with the crosshead displacement for all specimens up to the ultimate fracture loads which occurs in a dynamic way (Tong, 1998b). For lap shear joint specimens, the applied load increases linearly with the cross-head displacement until ultimate failure for these specimens with no crack. However, for the specimens with an embedded crack, it was noted that at least one load drop always took places prior to ultimate failure. In-situ microscopic inspection revealed that crack propagation and onset of delamination caused such load drops. Further increase in the controlled cross-head movement resulted in a further increase in the applied force to the specimen. For some specimens, several load drops happened before the ultimate failure happened with the lap shear joint being destroyed.

Table 4.2 lists the average ultimate failure loads and standard deviations as well as the associated failure modes for single lap specimens. Table 4.2 shows that the average failure loads of specimen SLJ-4, SLJ-6 and SLJ-8 are 11.1%, 9.2% and 5.5% lower that that of specimen SLJ-0. This result clearly indicates that (a) the embedded cracks in specimen SLJ-4, -6 and -8 can cause a noticeable reduction in the failure load; and (b) short cracks cause larger reduction in failure load than long cracks when the crack centres are placed right at the overlap ends. From this observation, it can be said that the adhesive bonded single lap joints tend to be tolerant to long cracks and to be less tolerant to short cracks. One reason for this is that the crack tip in the overlap is subjected to high peel and shear stresses when it nears the overlap end. The failure load of specimen SLJ-6-60 in Table 4.2 shows that the cracks embedded within the overlap can reduce the average failure load by 6.4% comparing to that of specimen SLJ-0. Finally, by comparing the average failure loads of specimen SLJ-0 and SLJ-6-40, it is noted that the embedded cracks don't seem to affect the ultimate failure load. This observation indicates that the joints are not sensitive to the embedded cracks located outside the joint overlap.

Table 4.2 Average failure load and mode of single lap joint specimens

Specimen group	Failure load (kN)	Failure mode type
SLJ-0	14.171±0.176	(a)
SLJ-4	12.598±0.606	(b)
SLJ-6	13.004±0.724	(b)
SLJ-8	13.395±0.627	(b)
SLJ-6-60	14.020±0.958	(a)
SLJ-6-40	13.263±0.360	(c)

Table 4.3 Average failure load, 1st load drop and displacement at failure, and failure modes of lap shear joint specimens

Specimen	Failure load (kN)	1st load drop (kN)	Failure displacement (mm)	Failure mode
LSJ-0	18.406±0.814	/	3.224±0.211	(d)
LSJ-2	12.682±0.794	0.586±0.327	2.382±0.123	(e)
LSJ-4	12.074±1.787	0.601±0.693	2.197±0.229	(e)
LSJ-8	11.594±0.812	0.357±0.123	2.104±0.296	(e)
LSJ-10	10.353±0.723	0.389±0.0.079	1.907±0.363	(e)

Table 4.3 lists the average initial failure loads, the average values of the first load drop, the average failure displacements and the failure modes observed. Evidently, presence of embedded crack causes a significant reduction in the average failure load, i.e. 31%, 34%, 37% and 44% reduction in the average failure load for these joints with a crack length of 2, 4, 8 and 10 mm respectively. The standard deviation for the failure load

is consistent of about 0.8 except for specimen LSJ-4 which is 1.787 and relative large. The results also indicate that the axial displacement at which failure occurs decreases as the crack length is enlarged from 0 to 10 mm.

For comparison, Figure 4.8 plot the average failure loads versus the crack length for both single lap and lap shear joints. It is observed that the lap shear joint configuration is more sensitive to the presence of the crack than single lap joints.

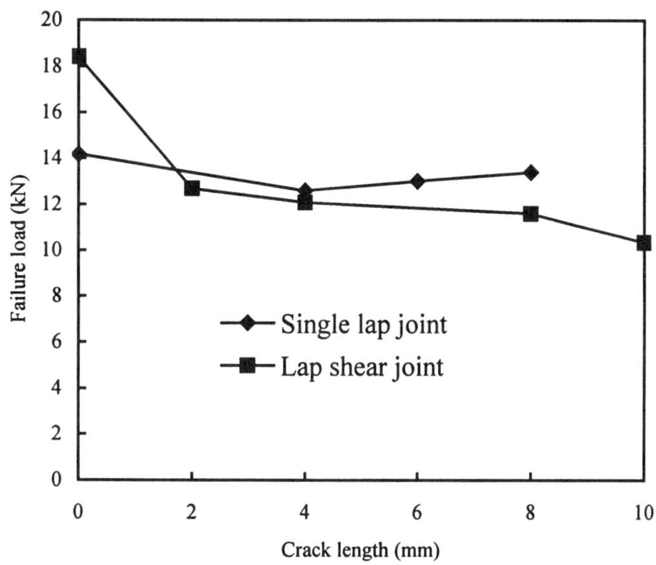

Figure 4.8 Effect of crack length on failure load of single lap and lap shear joint specimens

(b) Failure modes

Inspection of the failure surfaces of all specimens revealed that there were five typical failure modes for the two types of joint configurations. These five types of failure modes are referred as to failure mode (a), (b), (c), (d) and (e) respectively in the last column of Table 4.2 and 4.3, and are schematically depicted in Figures 4.9 and 4.10. The first three types of failure modes are observed in the fracture surface of the single lap joint specimens as detailed in Tong (1998b), while the last two are observed for the lap shear joints.

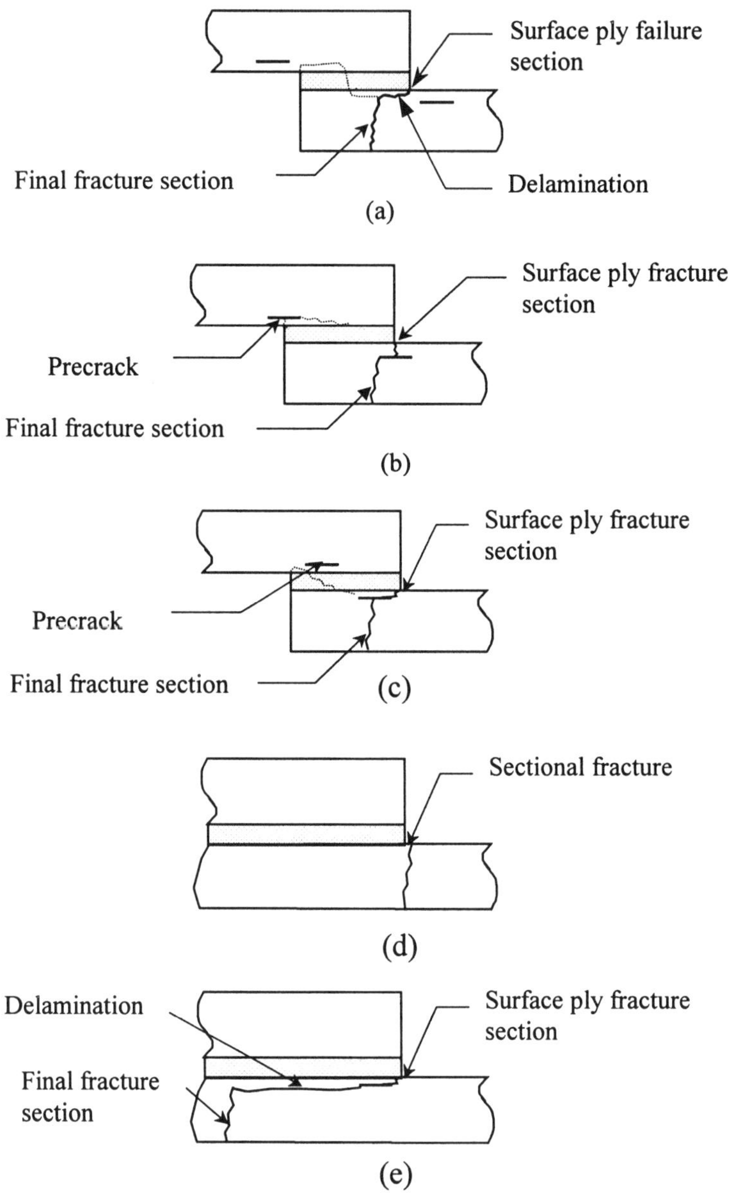

Figure 4.9 Typical fracture modes (a) specimen SLJ-0 A and SLJ-6-40, (b) specimen SLJ-4, SLJ-6 and SLJ-8, (c) specimen SLJ-6-60, (d) LSJ-0 and LSJ-2, LSJ-4, LSJ-8 and LSJ-10 (solid curves depicts typical cracks, and dash lines represents cracks for one or two specimens in the group)

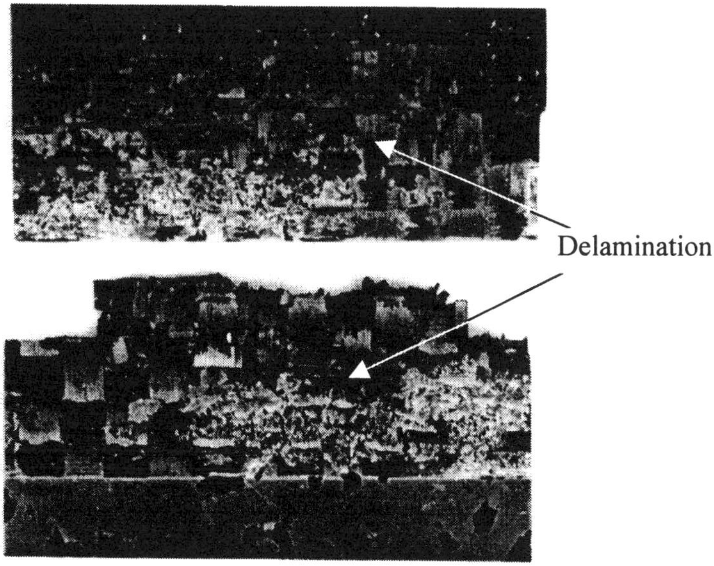

Figure 4.10a Typical fracture surface of single-lap joint specimen group A

Figure 4.10b Typical fracture surface of single-lap joint specimen group C (The white areas represent the Teflon film)

For most cracked specimens there was a surface ply fracture section and a final complete fracture of the remaining section. These two fracture sections are observed to be linked via an interlaminar delamination for most cracked joints or an intraply delamination for most virgin type of single lap joints (See Figure 4.10b). The delamination length varies and this indicates a non-uniform final complete section fracture, and thus implies a non-uniform crack propagation at the crack tips inside the overlap. A detailed discussion of these failure modes for each specimen group is given as:

Failure mode (A) (see Figure 4.9(a) and 4.10a): Initial sectional fracture due to high tensile stress in the surface ply develops into an intraply or interlaminar delamination and then leads to the final fracture of the remaining cross section.

Failure mode (B) (see Figure 4.9(b)): Initial sectional breakage occurs in the surface ply and then results in a final net sectional fracture near the crack tip in the overlap.

Failure mode (C) (see Figure 4.9(c) and 4.10b): Sectional fracture occurs in the surface ply at an overlap end and then develops into interlaminar or intraply delaminations that connect to the embedded crack near the crack tip. The final net sectional fracture happens at about half of the embedded crack length.

Failure mode (D)(see Figure 4.9(d)): The cross sections adjacent to the overlap end breaks completely for all virgin type lap shear joints. This represents a bending induced cross sectional fracture.

Failure mode (E) (see Figure 4.9(e)): This is similar to failure mode (b) but with a longer interlaminar delamination.

4.4 MODELING OF SINGLE LAP JOINTS

4.4.1 Finite element models

Two dimensional (2-D) plane-strain finite element models are used to predict the stress and strength for adhesive-bonded single lap joint specimens with and without cracks using Strand6 (G+D, 1995). The large deformation and rotation are included in the nonlinear analysis. The mechanical properties for the FM-300 film adhesive and a T300/934 cloth ply are given in Tables 4.4 and 4.5 (Anonymous; 1986, 1989).

Finite element models are generated to model all single lap specimen configurations with crack length of 4, 6 and 8 mm. In addition, an extra FEA model is generated to simulate a joint configuration with two 2 mm long cracks. In all calculations, 20 load increments are used to apply the load. Four-node quadrilateral plane strain elements are used to model both adherends and bondline. Figure 4.11 schematically depict the meshes used near the crack tips. As shown in Figure 4.11, small rectangular elements (0.0707 mm by 0.0667 mm) are used in the vicinities of the crack tips.

Table 4.4 Mechanical properties of FM300-K film adhesive (nominal bondline thickness 0.16 mm)

Young's modulus E (MPa)	2400
Poisson ratio μ	0.32

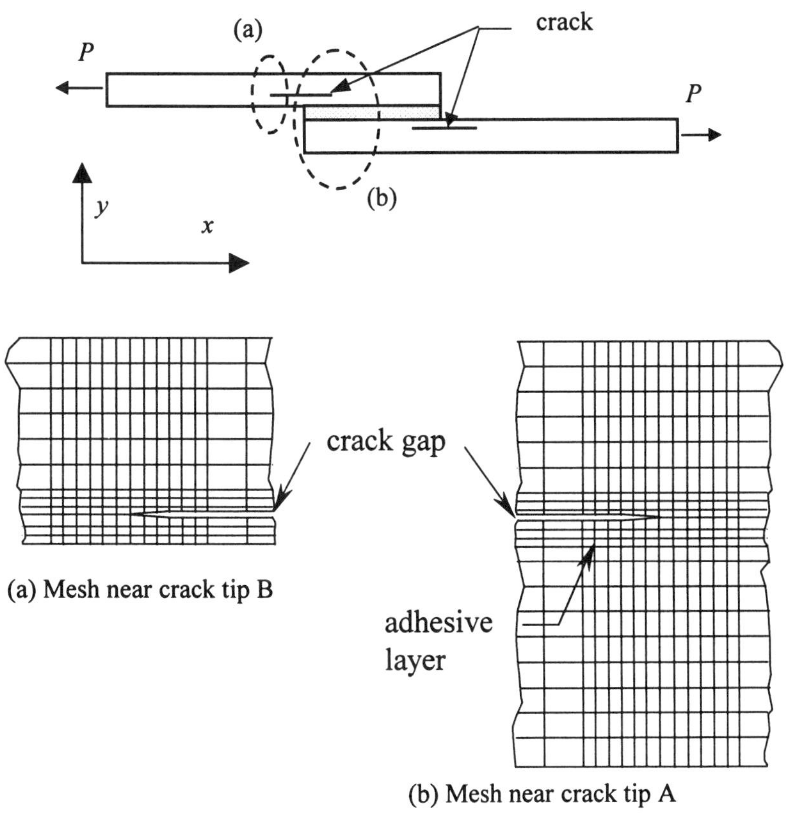

Figure 4.11 Sketches of the mesh used near the crack tips (Redraw from Tong, 1998b)

Table 4.5 Mechanical properties of a plain woven ply (nominal ply thickness 0.215 mm)

Longitudinal modulus E_1=E_2 (MPa)	57226
Transverse modulus E_3 (MPa)	4800
In-plane shear modulus G_{12} (MPa)	4481
Out-of-plane shear modulus G_{13}=G_{23} (MPa)	4400
In-plane Poisson ratio μ_{12}	0.05
Out-of-plane Poisson ratio μ_{13}=μ_{23}	0.28

4.4.2 Effect on stress distribution

The single lap joint modelled in Figure 4.11 is loaded with a final applied axial load P of 590 N/mm. Both ends of the joint are modelled as being fully clamped. Figures 4.12 and 4.13 show the distribution of the shear stress σ_{xy} and normal stress σ_{yy} in the adhesive layer for various single lap joints with different crack lengths. For the joints with cracks, the cracks are 2, 4, 6 and 8 mm long respectively, and are located between the first and second ply beneath the surface. Figures 4.12 and 4.13 show that the shear stress σ_{xy} and normal stress σ_{yy} in the adhesive layer in a joint with embedded cracks are different from those without a crack. Figure 4.12 reveals that, compared to the shear stress of a virgin joint, the shear stress in the adhesive layer of a cracked joint is smaller between the overlap end and the crack tip, and is larger between the overlap centre and a point slightly away from the crack tip. Figure 4.13 shows that the peak peel stress is reduced due to presence of the cracks, and this reduction is accompanied by appearance of another small peak peel stress. It is also observed that the reduction in peak peel stress is not sensitive to the length of the embedded crack.

Numerical results also show that stress concentrations exist at both ends of the overlap and the four crack tips as only purely elastic analysis has been conducted. Results of the different meshes indicate that the calculated stresses in the vicinities of the overlap ends and the crack tips are mesh dependent (Tong, 1998b). The singularities and mesh dependence exist even though an elastic-plastic analysis was carried out (Sheppard, Kelly and Tong, 1998). However, based on the finite element analysis results, the calculated stresses at crack tip A within the overlap are larger than those at crack tip B outside the overlap. This clearly correlates well with the experimentally observed failure mechanism that cracks propagated at crack tip A. This result will be confirmed further by comparing the values of the energy release rates at crack tip A and B.

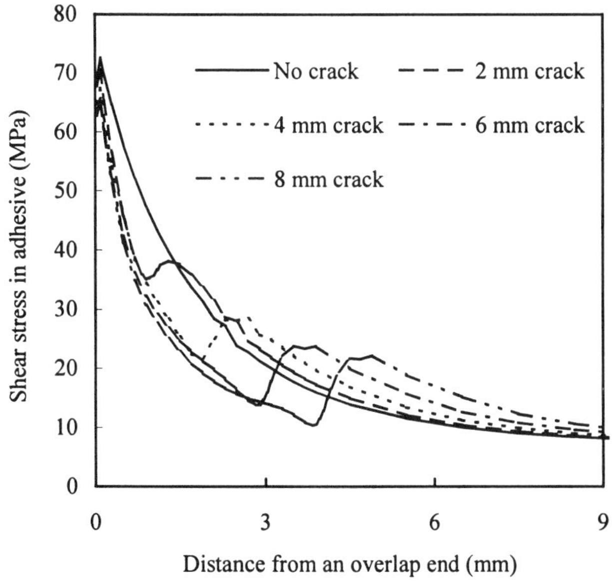

Figure 4.12 Effect of crack size on the shear stress distribution

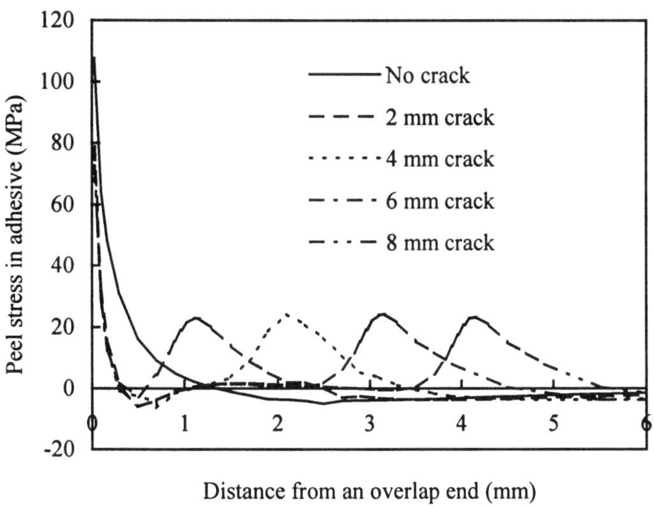

Figure 4.13 Effect of crack size on the peel stress distribution

4.4.3 Effect on energy release rates

The finite element analysis results for the joints with no crack and with cracks of 2, 4, 6 and 8 mm long are used to calculate the energy release rate at the crack tips A and B as shown in Figure 4.11. The mode I and mode II energy release rates at crack tip A and B are determined using the numerical procedure proposed by Rybicki and Kanninen (1977) as modified by Johnson (1987) (see Section 3.3). Figures 4.14 and 4.15 depict the energy release rates of mode I and II at crack tips B and A.

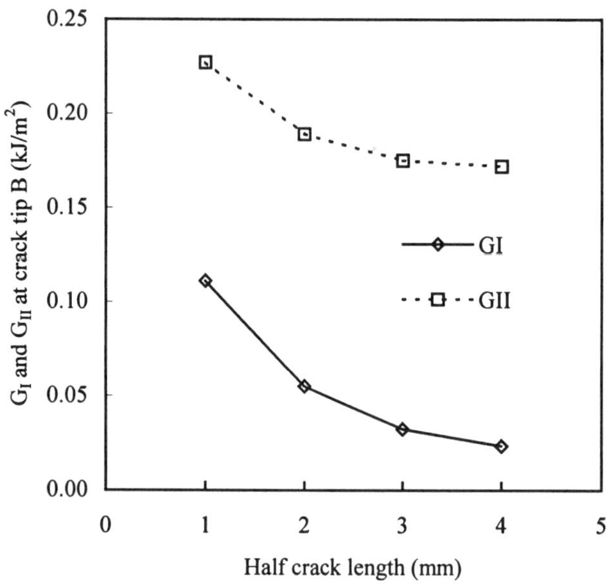

Figure 4.14 Energy release rate at crack tip B versus the half crack length

Figure 4.14 shows that the mode I and mode II energy release rates at crack tip B decrease as the crack becomes longer, i.e., the smaller the crack length, the larger the energy release rates. This indicates that crack propagation may initiate at a lower load for joints with a small crack length and at a higher load for joints with a large crack length. In addition, the mode II energy release rate is larger than the mode I energy release rate.

This means that crack growth at tip B may be controlled predominantly by mode II when the embedded crack is long. This could be the reason why the surface ply for some specimens breaks at the end of crack tip B.

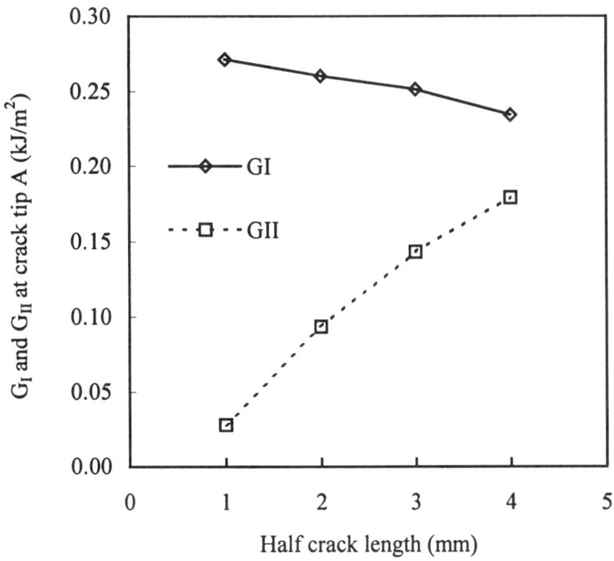

Figure 4.15 Energy release rate at crack tip A versus the half crack length

Figure 4.15 plots the energy release rates at crack tip A versus the half crack length. It is clearly shown that the mode I energy release rate at crack tip A decreases slowly and the mode II energy release rate increases when the crack length is increased. Unlike the case at crack tip B, the energy release rate of mode I is larger than that of mode II. The high peel stress overshoot near the crack tip A as shown in Figure 4.13 could be an contributing factor to the larger value of mode I energy release rate. For the material systems with the critical energy release rate of mode I being larger than that of mode II, crack propagation is mode I dominated and thus crack propagation can be expected to take place along the interface between the two plies. However, when the crack is long, the energy release rate of mode II increases, which in turn may cause the crack propagation to kink at an angle. This is in agreement with the observed crack kinking as shown in Figures 4.9 and 4.10.

Figure 4.14 and 4.15 also show that the mode I energy release rate at crack tip A is significantly larger than that at crack tip B, while the mode II energy release rate at crack tip A is smaller than that at crack tip B. For the material system tested, the critical value of mode I energy release rate is much smaller than that of mode II energy release rate, thus crack tip A is more critical than crack tip B for all the cracked joints. This probably explains why crack propagation was experimentally observed only at crack tip A.

The calculated results shown in Figure 4.14 and 4.15 may also be used to interpret the scenario of the failure sequence for adhesive bonded composite joints as postulated by Hart-Smith (1973a). The scenario was schematically depicted in Figure 3.4 for a double lap joint. In this scenario, small delamination between the first and second ply occurs due to induced high peel and shear stresses at the interface between the two plies. Immediately after onset of the small delamination, the crack at the tip within the overlap (crack tip A) grows towards the middle of the overlap, due to the high value of energy release rate of mode I.

4.4.4 Effect on joint strength

The critical energy release rates of mode I and II for the T300/934 plain weave material are G_{IC}=0.23 kJ/m^2 and G_{IIC}=0.66 kJ/m^2 (Tong, 1998b; Saghizaden and Dharan, 1986). The strength of the cracked joints can be predicted using the following simple bilinear interaction criterion (Reeder, 1993) as given in section 3.3, i.e. the Energy Release rate based Criterion (ERC):

$$\frac{G_{II}}{G_{IIC}} + \frac{G_I}{G_{IC}} = 1$$

The strength of the joints can also be predicted using the Maximum Stress Criterion (MSC) over the surface ply, namely, it is assumed that failure occurs when the average tensile stress in the surface ply attains its in-plane tensile strength. The in-plane tensile strength is 518 MPa which was measured following the ASTM standard.

Figure 4.16 compares the measured average failure loads and those predicted using the above energy release rate based criterion (ERC) and the maximum stress criterion (MSC). A reasonable good correlation is observed between the measured and predicted failure loads. However, the finite element analysis predicts that the failure loads attains a plateau when the crack is long while the average measured failure loads increases back towards that of the virgin joints.

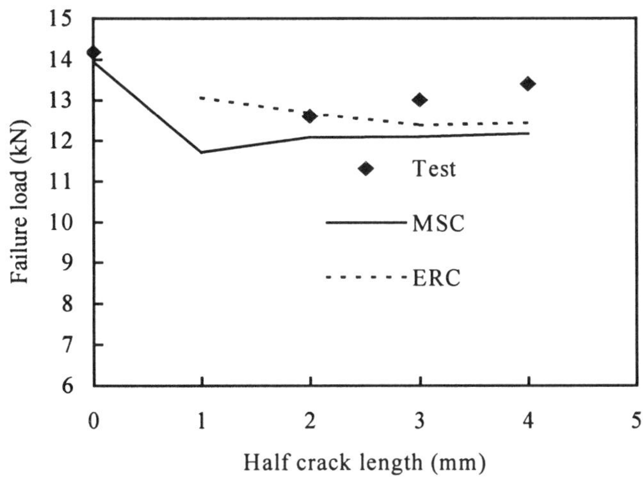

Figure 4.16 Comparison of the average measured failure loads and those predicted using finite element analysis

4.5 AN ANALYTICAL MODEL

When a crack exists between two adjacent plies in an adherend (see Figure 4.6), its presence significantly redistributes the stresses in the adhesive layer as shown in Figure 4.12 and 4.13, and thus makes it more difficult and challenging to obtain an analytical solution. Tong (1998) presented a solution procedure based on a decoupled global and local analysis for the adhesive bonded single lap joint shown in Figure 4.6(b).

In the global analysis, the peel and shear stresses in the adhesive are not included. The global solutions for the deflections and stress resultants for a virgin type of composite single lap joint are obtained following the solution procedure presented by Goland and Reissner (1944). The global bending moments M_A and M_B and the shear forces Q_A and Q_B at the crack tip A and B can be obtained.

In the local analysis, the joint is divided into a number of segments. As shown in Figure 4.17 the half of the joint can be divided into three segment, CB, BA and AD. The cracked portion BA of the joint can be further divided into an upper and a lower beam portion. The upper beam portion has a uniform cross section while the lower beam portion has a variable cross section comprising different materials. The axial force, the bending moment and the transverse shear force at the ends of upper and lower portion of the cracked joint are then determined by imposing continuity conditions at both ends of the lower and upper beam parts.

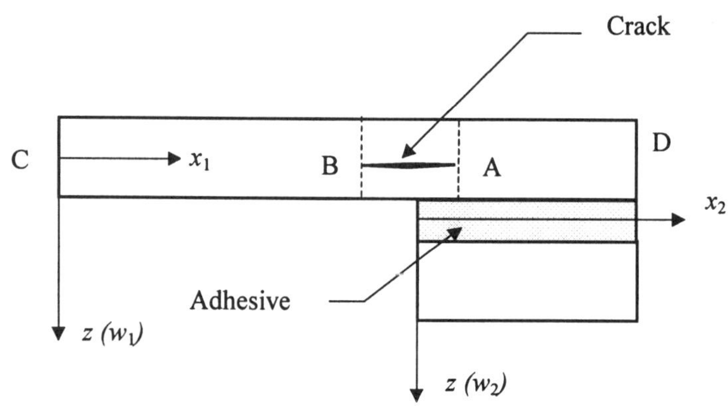

Figure 4.17 various segments in half of a single lap joint with an embedded crack

The axial force in the lower beam portion at crack tip B are calculated using both the finite element analysis and the analytical solution methods for joints with a crack length of 2, 4, 6 and 8 mm and subject to an axial load of 590 N/mm. Figure 4.18 shows the comparison between the axial force predicted by using the analytical and the finite element analysis method. There is a reasonably good agreement between the FEA results and the analytical predictions for the crack length of 2 to 8 mm. Thus the analytical model can predict the axial forces with a good accuracy of less than 12%.

The axial force can be used to calculate the maximum stress in the surface ply for the joints with various crack lengths. By applying the maximum stress criterion over the surface ply, the joint strength can be obtained. The predicted joint strengths based on the analytical solution procedure are plotted in Figure 4.19 with the average measured failure loads, and a reasonable good correlation is observed. Again the recovery in failure load with increasing crack length is to be noted. This aspect can be regarded as being beneficial for damage tolerance.

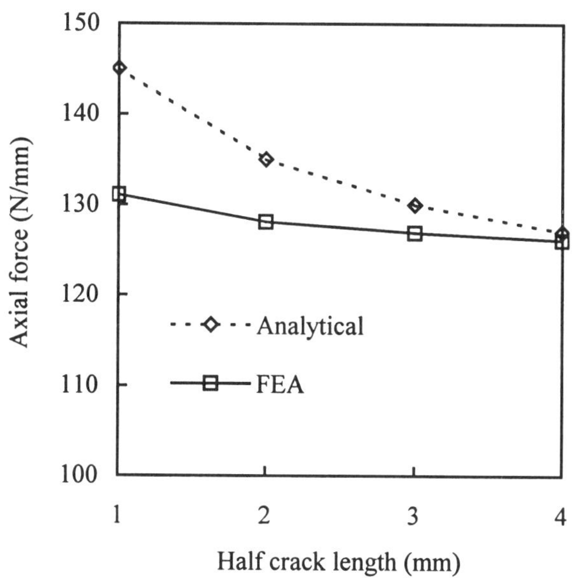

Figure 4.18 the calculated axial force in the lower beam portion at crack tip B versus the half-crack length

The stress resultants in the upper and lower beam sections at both crack tips (A and B) can be used to calculate the energy release rates in conjunction with the method suggested by Hutchison and Suo (1992). However, using this a poor correlation is obtained between the analytical and the finite element analysis results (Tong, 1998b). As pointed out by Tong (1998b), it may not be appropriate to model the upper and lower adherends near the crack tips using classical beam theory. One remedy to this is to apply a higher order beam theory rather than the classical beam theory. Another contributing factor could be due to the fact that the global and local analyses are decoupled. A coupled analysis may be required to improve the poor correlation in energy release rate.

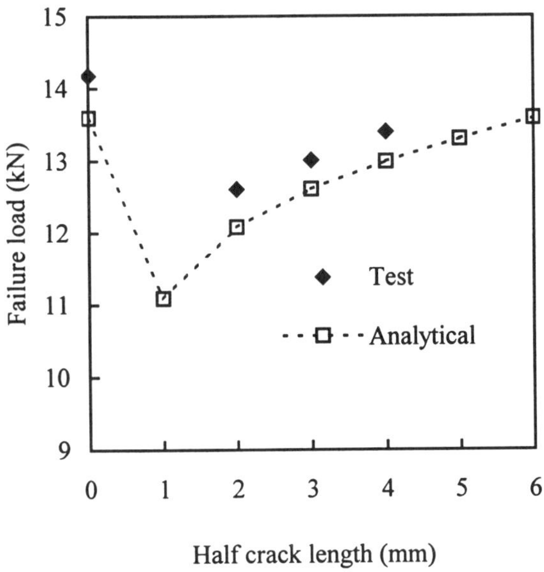

Figure 4.19 Failure loads measured and predicted using analytical solution versus the half-crack length

Figure 4.20 shows the effect of the crack depth from the surface on the failure load predicted based on the analytical solution for various half-crack lengths. The 0.215 mm depth represents the crack between the first and second plies while 0.43 mm depth represents the crack between the second and third plies. This figure clearly demonstrates that a short crack between the first and second plies near the adhesive interface is more detrimental than a long one far from the adhesive interface between the second and third plies.

4.6 SUMMARY

Co-cured and secondary bonded composite structures are prone to accidental impact damage. Such impact damage, especially the low energy impact damage, can significantly reduce the load carrying capacity and thus seriously affect structural integrity and durability (Springer and Fin, 1991; Sierakowski and Newaz, 1995). This chapter shows that when damage occurs in the bonded area near the overlap end the load carrying capacity of the bonded structure can be significantly affected. For the lap shear joint, about 40% reduction in the measured failure load is observed.

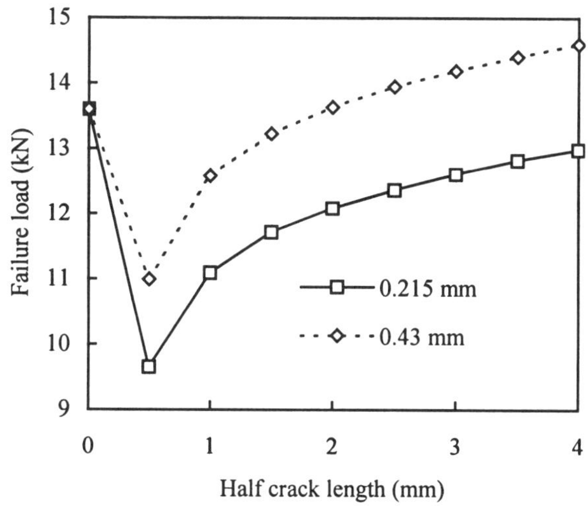

Figure 4.20 Effect of crack depth on the predicted failure loads

There exist an enormous amount of research directed towards understanding the effect of through-the-thickness cracks in sheet in bonded stiffened structures. However, only a very limited amount of work on the effect of delamination, especially near bonding area, on the structural performance of bonded stiffened structures has been undertaken. Efficient and effective methodologies are needed to evaluate the effect of delamination on structural integrity and durability of bonded stiffened structures. For the delaminated lap joints studied in this chapter, a number of questions are still yet to be answered. For example, a simple coupled global and local analysis model needs to be developed. The effect of loading types, e.g. compression and bending, also needs to be studied. The ultimate goal has yet to be achieved in understanding the effect of multi-layer interfacial cracks and true impact damage on structural strength and fatigue life of secondary bonded stiffened plate and shell type structures.

5

Optimum design

5.1 INTRODUCTION

As an engineer it is one task to design a structure which performs to a specified adequate level and it is an even harder task to design a structure that is as perfect as possible. If the task were energy conversion then 100% electrical energy conversion to mechanical energy in an electric motor would be perfect. If the task were aerodynamic design then zero drag with maximum lift would be perfection. If the task were the structural design of an adhesive joint then what would be the conditions for a perfect joint; maximum load capacity, minimum weight and maximum fatigue life? For a structure in general there are many requirements for a good/perfect life, however for an adhesive joint the list is restricted to strength, weight and failure mechanism.

Optimal design of an adhesive joint must be based on (a) the nature of the materials to be joined, (b) the method of joining, and (c) the strength and weight analysis. For instance, it is not sufficient to simply replace riveting, bolting or welding with adhesive bonding. No matter what forms of joints are used in a structure, they are potentially the weak locations in the structure with weight penalty. Hence it is always desirable to optimally design a structural joint to achieve its intended functions as perfectly as possible. When optimizing a structural bonded joint, a designer should consider the following aspects:

(a) Service conditions, such as stresses and environmental conditions, which are likely to be encountered in service,
(b) Selection of material combinations, such as selection of suitable adhesive and adherends, and
(c) Manufacturing specifications, such as surface pretreatment, fabrication procedures, quality assurance and protection from unacceptable hostile conditions in service.

It is apparently desirable to optimally choose the relevant design parameters so that optimum performance of the structural bonded joint can be achieved. In one aspect, a designer may pursue optimum combinations of adhesive and adherends, optimum surface treatment parameters, and optimum manufacturing procedures in order to optimize the joint performance, e.g. durability, resistance to hostile environmental conditions. In another aspect, a designer may pursue an optimal joint geometry with a maximum static strength and fatigue life and minimum weight.

The strength requirement clearly dictates that under all operational environments there is no deterioration in normal operational conditions and that under abnormal conditions there is regulated deterioration in performance. Failure occurs only at the design limit load. The weight optimality criterion dictates that the joint is of as low a weight as possible and that excess material is to be avoided. As a measure of this weight criterion the concept of a neutral joint could be introduced whereby the presence of a joint, from the perspectives of stiffness, strength and weight, is compared with what would have happened had there been no joint at all. This is a similar concept to the neutral hole whereby a perimeter-reinforced hole does not raise stress and mass over what would have been present had there been no hole.

The chapter starts with some traditional calculus based optimization on a double strap joint using 1-D shear lag theory. In the same theoretical framework the inverse problem of a fully stressed joint is considered. With such preliminaries over some general remarks are made on methods of structural optimization and how such methods can be applied to the optimization of joints. From these methods two are selected as appropriate for shape optimization of adhesive joints, namely; shape vector adjustment and a relatively new method called Evolutionary Structural Optimization (ESO) which makes shape optimization for adherends a particularly simple task. One example is presented using shape vector adjustment and two examples are presented using the ESO software called EVOLVE which is available from the authors at no cost.

5.2 1-D LINEAR OPTIMUM ANALYSIS

5.2.1 Double strap butt joint

Before introducing the broad topic of joint optimization, it transpires that a lot can be gained from undertaking simple calculus based optimization on a

1-D double strap joint. This is a good example to use, as the co-linearity of the end load vectors reduces the real effect of joint rotation and thus the analysis is closer to reality than a single lap joint where rotation is inevitable. Figure 5.1 shows the outer adherends of a material with Young's Modulus E_1 with thickness t_1 and tension T_1 per unit width whilst the displacement is denoted by u_1, and the central adherend of a material with Young's Modulus E_2 with thickness t_2 and tension T_2 per unit width whilst the displacement is denoted by u_2. The adhesive is of thickness η and has a shear modulus of G and its shear stress is denoted by τ

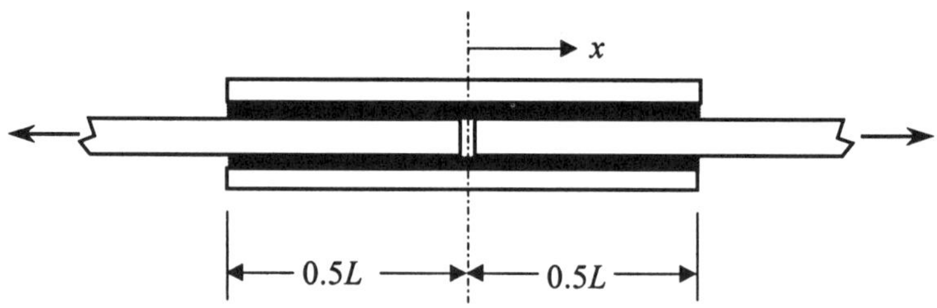

Figure 5.1 Schematic of double strap butt adhesive joint

The stress resultants, the shear stress and the displacements are all functions of x shown in Figure 5.1. These central adherends are not connected ensuring that the load has to diffuse into the straps and back again.

It is assumed that no significant bending takes place in the joint, and it is also assumed that shear stress is constant across the bondline thickness. Based on the assumptions, the stress/strain/displacement relationships are expressed for the adherends and the adhesive as

$$T_1 = (Et)_1 \frac{du_1}{dx}, T_2 = (Et)_2 \frac{du_2}{dx}, \quad \tau = \frac{G}{\eta}(u_1 - u_2) \tag{5.1}$$

and the equilibrium equations are

$$\frac{dT_1}{dx} = \tau, \qquad \frac{dT_2}{dx} = -2\tau, \qquad P = 2T_1 + T_2 \tag{5.2}$$

After eliminating the displacements, we have

$$\frac{d^2T_1}{dx^2} - \alpha^2 T_1 = -\frac{P}{D} \tag{5.3}$$

where

$$\alpha^2 = \frac{G}{\eta}[\frac{1}{(Et)_1} + \frac{2}{(Et)_2}] \qquad D = \frac{(Et)_2 \eta \alpha^2}{G} \tag{5.4}$$

The solution to (5.3a) can be expressed as

$$T_1 = C_1 \cosh \alpha x + C_2 \sinh \alpha x + \frac{P}{D} \tag{5.5}$$

In which C_1 and C_2 are constants of integration. The appropriate boundary conditions for the end tensions are

$$\begin{aligned} T_1 &= 0 \quad at\ x = 0.5L \\ T_2 &= 0 \quad at\ x = 0 \end{aligned} \tag{5.6}$$

Solving these for the constants gives the analytical expressions for the adherend tension and the adhesive shear stress as follows,

$$\begin{aligned} T_1 &= PH \cosh \alpha x - \frac{\sinh \alpha x}{\sinh(0.5\alpha L)}\left[\frac{P}{D} + PH\cosh(0.5\alpha L)\right] + \frac{P}{D} \\ \tau &= PH\alpha \sinh \alpha x - \frac{\alpha \cosh \alpha x}{\sinh(0.5\alpha L)}\left[\frac{P}{D} + PH\cosh(0.5\alpha L)\right] \end{aligned} \tag{5.7}$$

where $H = 0.5 - 1/D$. The variation in shear stress has the typical dip in the middle with upturns at both ends as presented in Chapter 2.

The purpose of optimization is to maximise the load transfer through the joint for a given limiting shear stress in the adhesive and to make the stress distribution as iso-stressed as possible. Considering the second objective first, we could begin by asking that the shear stress at both $x=0$ and $x=L/2$ be equal. After a lengthy of manipulation, this requirement reduces to

$$HD = 1 \qquad \text{or} \qquad \cosh(0.5\alpha L) = 1 \tag{5.8}$$

After further substitution, the first of these conditions reduces to

$$(Et)_2 = 2(Et)_1 \tag{5.9}$$

This tells us that the extensional stiffness of the centre adherend has to be equal to the sum of the two straps. This is what could be referred to as a balanced joint. On reflection this result is a logical one. The second condition for equalisation of the shear stress end values which tells us

$$L \to 0 \quad or \quad \alpha \to 0, \quad ie.\, \alpha^2 = \frac{G}{\eta}\left[\frac{1}{(Et)_1} + \frac{2}{(Et)_2}\right] \to 0 \tag{5.10}$$

Thus if the length of the joint tends to zero we achieve a balanced joint. It will be shown later that if this design path is followed that it also has the effect of making the shear stress distribution tend to be a constant. Initially this would appear to be a beneficial result however the load capacity of the joint goes to zero as the length reduces. The second of the conditions can never be physically achieved but does give an indication of the direction in which material and geometric parameters need to change to balance the joint. If the adherend and adhesive thicknesses and the Young's Modulus of adherend get bigger or the adhesive shear modulus gets smaller the end joint shears will move closer together.

If we assume that the joint is now balanced and consider the relationship between the maximum allowable adhesive shear stress, call it Y and the load P transmitted through the joint

$$Y = \frac{\alpha P}{4\sinh(0.5\alpha L)}\left[1 + \cosh(0.5\alpha L)\right] \tag{5.11}$$

or rewritten as

$$\frac{P\alpha}{2Y} = \frac{2\sinh(0.5\alpha L)}{[1 + \cosh(0.5\alpha L)]} \tag{5.12}$$

It is of interest to designers to examine the maximum load carrying capacity of joints, plotting $(P\alpha/2Y)$ versus $(\alpha L/2)$ in Figure 5.2 shows that there exists an asymptote. It is interesting to note that the maximum load is in independent of length when $\alpha L>10$. However, a sufficient overlap length is required to ensure a large elastic trough for fatigue resistance (Hart-Smith, 1973a).

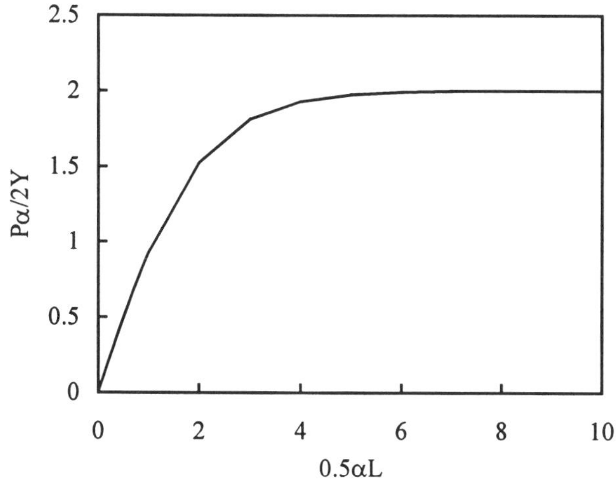

Figure 5.2 Plot of $P\alpha/2Y$ versus $0.5\alpha L$.

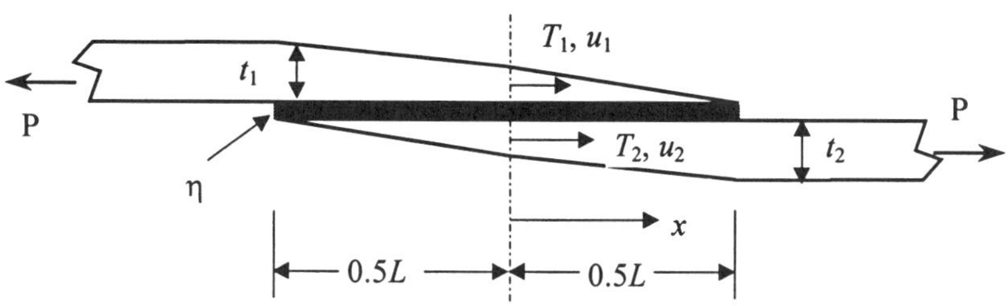

Figure 5.3 Single lap joint with variable thickness adherends

5.2.2 A scarf single lap joint

The derivation of the small deflection linear analysis of the single lap joint follows the same basic procedure as the double butt strap joint of section 5.2.1. Figure 5.3 shows the upper adherend of a material with Young's Modulus E with thickness t_1 and tension T_1 per unit width whilst the displacement is denoted by u_1. The subscript 2 is used for the bottom adherend. The adhesive is of thickness η and has a shear modulus of G and its shear stress is denoted by τ. The stress resultants, the adherend thickness and the displacements are all functions of x only.

The stress or force and displacement relationships are identical to equation (5.1), while the equilibrium equations are

$$\frac{dT_1}{dx} = \tau, \quad \frac{dT_2}{dx} = -\tau, \quad P = T_1 + T_2 \tag{5.13}$$

With these basic structural mechanics equations and we can start with the assumption that the shear stress is constant, from the first of (5.13)

$$T_1 = \tau x + C \tag{5.14}$$

Using the boundary conditions

$$T_1 = P \quad \text{at } x = -0.5L, \qquad T_1 = 0 \ \text{ at } \ x = +0.5L \tag{5.15}$$

These give

$$C = \frac{P}{2} \quad \Rightarrow \tau = -\frac{P}{L}, \quad T_1 = P(\frac{1}{2} - \frac{x}{L}), \quad T_2 = P(\frac{1}{2} + \frac{x}{L}) \tag{5.16}$$

Using the 3rd equation in equation (5.1) and the 2nd equation in the above equation, we find

$$\frac{du_1}{dx} = \frac{du_2}{dx} \tag{5.17}$$

Now using the first two equations in equation (5.1) and assuming both adherends made of the same material, we have

$$\frac{(Et)_1 \dfrac{du_1}{dx}}{(Et)_2 \dfrac{du_2}{dx}} = \frac{t_1}{t_2} = \frac{T_1}{T_{2\,1}} = \frac{(\dfrac{1}{2} - \dfrac{x}{L})}{(\dfrac{1}{2} + \dfrac{x}{L})} \tag{5.18}$$

Putting in the constant for this ratio, given that the initial adherend thicknesses are t_0 at the start and end of the joint overlap, namely, $x=\pm 0.5L$, gives

$$t_1 = t_0(\frac{1}{2} - \frac{x}{L}), \quad t_2 = t_0(\frac{1}{2} + \frac{x}{L}) \tag{5.19}$$

This shows that for this simplified analysis, the 'optimum joint' has a linear taper with the shear stress in the adhesive constant. An ideal situation that later analysis will question. This simplified analysis on an optimum fully stressed single lap joint has generated several derivative joint configurations including bevelled joints, sloping joints and multi-step lap joints. These derivative forms of structural adhesive joints have the attribute of shaping the adherends to obtain a more favourable stress situation.

The simplified analyses used in these two optimization examples presented above, yield certain design observations that are used in current practice. The need to get a balanced joint where the shear stress peaks are equalized and the need to get a set of geometric and material parameters that maximise the load transfer for a given adhesive failure stress have led to such above mentioned design observations. However despite the prevalence of designs based on the simple 1-D analysis, the fact that it is incapable of accommodating peel stress is a major drawback. The FEA models shown in Chapter 2, even using a small displacement analysis with no geometric non-linearity all have peel stresses, often of the same magnitude as the shear stress. Thus methods need to be established that can optimize a joint material and geometry that include realistic effects. The following section is an overview of currently available methods of structural optimization and some comments upon their applicability to joint optimization.

5.3 METHODS OF STRUCTURAL OPTIMIZATION

Structural optimization is a fusion in the areas of engineering, mathematics, science and technology that has the goal of achieving the best performance for a structure, be it a bridge or a space vehicle or a spectacle frame. All aspects of the environment the structure lives in must be accommodated even undesirable ones, crash worthiness of cars being one example and one more relevant to this text is barely visible impact damage in the vicinity of a bonded joint that may be damage intolerant.

Because of its mathematical complexities, the topic of structural optimization remained of more academic interest until the last twenty years. Since then there has been a re-focusing on the topic. This is being pulled by the large number of highly competent analysts who are skilled in Finite Element Analysis (FEA) and who now want to use the method beyond the previous 'checking designs out' capability. At the same time, two push factors, have changed the topic from the previous narrowness to the stage where it is becoming accessible to the practising engineers and scientists. First the availability of high performance computing power at low cost, and secondly the rapid improvements in algorithms used for design optimization where thousands of design variables and constraints may need to be handled. There have been more and more research activities directed towards making these developments available to engineers and scientists in an easy, reliable, inexpensive and understandable form.

Several classes of methods of structural optimization are proving useful in this emerging role. The classical calculus-based design parameter optimization methods have become more efficient due to improvements in algorithms used in the search engines (Vanderplaats, 1984; Haftka and Gurdal, 1992). The next powerful class of methods are generically grouped together with the title Optimality Criteria (Rozvany; 1989). More recently the Homogenization Method, pioneered by Bendsoe (1995) has proven to be successful in generating optimum topologies for continuum structures by treating element densities as design variables. Even most recently the Evolutionary Structural Optimization (ESO) method has been investigated and developed by Xie and Steven (1997).

There are three distinct forms of structural optimization with different solution strategy. In a given design situation all three can be present.

5.3.1 Topology optimization

Topology optimization exists where the actual form of the structure is unknown in advance. The designer does know all the environments the structure has to live in and the optimality criteria that need to be applied. In Aerospace this could be the shape of the wing carry through box for a large civilian aircraft that also has under-carriage pick up points. There are several load cases for all the different flight regimes and several different sets of support conditions for in-flight and landing. The optimality criterions are most likely to be on stress and stiffness with possible subsidiary ones on frequency and buckling. Some of this describes the so-called MBB beam, originally proposed by Airbus, which has been the subject of many optimization studies (see, Olhoff et al, 1991; Xie and Steven, 1997).

The object of topology optimization is to have no restriction on the final form of the structure. It is like starting from a blank, but full, sheet of paper and then drawing a topology that best meets the criteria. A traditional single criterion where there is only one load case has been the fully stressed design (FSD) where each part of the final structure is at the same stress. Whilst this seems attractive, a little reflection will reveal that when there are multiple load cases and support cases there is no way a fully stressed design can be achieved. What is best, is some form of weighted compromise, mathematically the weights turn out to be the Lagrange Multipliers associated with the degree of satisfaction of the individual criteria with the Kuhn-Tucker conditions (Vanderplaats; 1984) applied to those inactive constraints at the optimum point.

There are currently three methods that are effective in solving topology optimization problems, and one emerging method, these are, in historical order of appearance;

Optimality Criteria: Here elements in a finite element mesh are altered or removed by applying optimality criteria derived by applying the duel theory of convex programming to a separable approximation of the design problem (Rozvany et al, 1995).

Homogenization: Here the microstructural form of the material has different shapes governed by two or more local parameters. For each element in the FE mesh these parameters become the design variables and mathematical programming techniques with sequential quadratic programming used to determine the optimal material distribution. The technique was pioneered by Bendsoe (1995).

Evolutionary Structural Optimization: Here a fine FE mesh is constructed and after subsequent analysis elements that are lowly stressed/not contributing/ineffective are slowly removed. The evolution rate is low and often many 100's of FE calculations are needed. The concept has been around for many years but since 1993, Xie and Steven (1997) have made detailed studies and established rules that make the method work for the full range of structural situations.

Genetic Algorithms (GA): More recently these methods have had limited success in solving structural optimization problems. Genetic algorithms are based on the theory of natural selection where the structural topology optimization can be achieved by successive ranking of populations of elements and elimination the weakest (Goldberg, 1989).

5.3.2 Shape optimization

With shape optimization the topology of the structure is known and there is some aspect of detail that is giving rise to a high stress, such as a fillet or a notch. The object of shape optimization is to find the best shape that will have the best stress outcome. When there is only one load case then shape optimization is equivalent to a Mini-Max problem where the maximum stress on the surface of the region is minimized. In other words the shape of the local surface is altered till the maximum stress is as small as possible. Further reflection will demonstrate that this is equivalent to the surface becoming an equi-stress contour. When there are several load cases or support cases then this is no longer the case and the best that can be said is that the shape is the best compromise.

A typical situation where shape optimization is present is the shapes of forks in trees, the shape of a fillet between two sections of different width. These are external surfaces. Indeed any of the geometries where the topic of stress concentrations is present are suitable for consideration using shape optimization (Haftka and Grandhi, 1985; Ding, 1986). When surfaces are internal the shape optimization process will find the best shape for the internal cut out, e.g. single or multiple holes in composites (Falzon et al 1996).

As with topology optimization there are several techniques available for solving the shape optimization problem, these are listed below.

Mathematical programming. The shape is described mathematically by a series of parameters (design variables) and the objective function is described in terms of these variables. The differentials of the objective with respect to the variables are obtained directly or by computation with a finite difference form for the differential. Second differentials are obtained for the Hessian matrix and conjugate gradient, steepest decent or quadratic programming search engine used to find the set of design variables that fit the design criteria (Pederson et al., 1992; Kristensen and Madsen, 1972).

Evolutionary Structural Optimization. The same basic algorithm as is used for the topology optimization form of ESO is modified so that only surface element in the design region are available for removal, this is called nibbling. The effect of this is to undertake shape optimization. The advantage is that there can be multiple environments (Xie and Steven; 1997).

Simulated Biological Growth. This techniques pioneered by Mattheck and Burkhardt (1990) in Germany uses a gambit whereby stress at the surface being shape optimized is converted to temperature and the surface allowed to move outwards, if stress high, or inwards if stress low. The motion is achieved by the thermal expansion equivalent of the fake temperature. Several iterations are needed. The method produces good results but, to date, only for a single load case.

5.3.3 Size optimization

The structure is defined by a series of sizes and dimensions. Combinations of these sizes and dimensions are sought that achieve the optimization criteria. There are two major categories of problems in size optimization namely; discrete structures and continuum structures.

Discrete structures include pin and rigid jointed structures. This is the area that has received to most attention over the last forty years. A structure is defined and its loads and supports. The cross sectional dimensions of the members are adjusted according to the optimization goal(s). Traditionally mathematical programming techniques and the optimality criteria techniques have been used successfully. If member dimensions can go to zero then they are unnecessary and a further reduced structure can be produced, this situation is sometimes called layout optimization. Several examples are given in the text by Xie and Steven (1997).

The optimization of continuum structures includes, aircraft style structures such as carbon fibre laminates, stiffened panels, wing layouts, with spars and ribs where the structure can be describes as series of sizes or parameters, such at stiffener pitch, skin thickness, ply angle. Optimization techniques are then used to find the combination of design variables that give the best result. Methods generally used are mathematical programming and genetic algorithms. Also in this category of size optimization with continuum structures is thickness variation of plates and shells.

5.4 RESEARCH ON SHAPE OPTIMIZATION OF BONDED JOINTS

Joints are sources of stress concentrations that diminish the overall efficiency of a structure. In strength critical components it becomes imperative to reduce these stress concentration factors so as to increase structural efficiency. As demonstrated in Chapter 2, stress concentrations in bonded joints arise from abrupt changes in adherend/adhesive thickness and from differences in elastic modulus. Limited work can be found to date in the use of FEA optimization on adhesive bonded joints.

Lerchenthal (1964) employed some simplifying assumptions and determined the shape of adherends, which produce as far as possible a uniform shear stress distribution in the direction of the bondline. Photoelastic techniques were also used to verify shear stresses in the bondline.

Cherry and Harrison (1970) postulated that the objective to obtain an optimum single lap joint should be that of achieving a uniform shear stress along the overlap length. A closed form solution was employed to this effect resulting in a uniformly tapered joint.

Ojalvo (1985) studied a procedure to minimize elastic shear stress concentrations in adhesive lap joints. The procedure consisted of employing an inverse method which imposes a desired shear stress distribution and then determines the required adherend shapes to achieve it. The objective function in this case was based on shear stresses, thus the solution obtained by this method needed to be checked to determine the actual shear and peel stresses in the joint. Optimum adherend shapes presented for both single and double lap joints are single curvature tapers.

Groth and Nordlund (1991) studied optimum shapes of adherends with the intention of minimizing the Tsai-Hill function along thc adhcsive overlap. The Tsai-Hill yield criterion is a general form of the Von Mises yield hypothesis for orthotropic materials. Joint types included in the analysis were single lap, double lap, double-strap and a console bonded to a rigid wall. The adherends for both the single and double lap joints had to be pre-profiled so that the mesh distortions would not make the analysis unfeasible. The slave and master variables were employed to enforce anti-symmetry for the single lap joint case. The explicit linear constraints had to be employed to ensure that the finite element mesh did not fail during the optimization process. The optimum shape found for the adherends of single lap joints by Groth and Nordlund (1991) consisted of double tapered adherends with a flat region in between both tapers.

In the next section, two different optimization methods are used for the shape optimization of adherends and adhesive fillets in an attempt to correlate optimum shapes achieved by employing different approaches (Rispler, 1998). The first method is the shape basis vector technique (Moore, 1994). This method employs approximation concepts and mathematical programming techniques to achieve optimum shapes of structures given a design objective and constraints. The second method is the evolutionary structural optimization method (ESO) proposed by Xie and Steven (1997). This method does not require any mathematical formulations or the reinforcing of constraints. It also does not require the use of an objective function. The ESO method is broadly classified as a heuristic optimality criteria method.

5.5 APPLICATION OF SHAPE BASIS VECTOR TECHNIQUE

The shape basis vector (SBV) approach (Moore, 1994) is a fairly recent method which employs mathematical programming techniques for shape optimization of parts. This method has been adopted by the finite element analysis code MSC/NASTRAN within their optimization solution sequence (SOL 200). This module allows the geometry of the design model to be varied to improve a design objective (e.g. minimum weight) while satisfying various design constraints (e.g. maximum stress). The variation of geometry is achieved in the MSC/NASTRAN by changing the locations of nodes. The coordinates of these nodes are controlled by design variables while their movement is represented by shape design variables. A shape basis vector is a group of design variables. This vector contains information on the magnitude and direction of shape variation. In a finite element model, the geometric shape is expressed in terms of node coordinates. The following equations outline the relationship between the new and old geometry:

$$\{G\}^{new} = \{G\}^{old} + \sum_{i=1}^{n} \left(X_i^{new} - X_i^{old}\right)\{T_i\} \tag{5.20}$$

$$\{\Delta G\} = \sum_{i=1}^{n} \Delta X_i \{T_i\} \tag{5.21}$$

where $\{G\}$ is the node coordinate vector, $\{T\}$ is the shape basis vector, X_i is the design variables, n is the number of design variables, the superscripts *new* and *old* are at the end and beginning of design cycle.

Equation (5.21) shows that the new shape is linearly dependent on a combination of the shape basis vector and the *old* shape. The shape basis vectors are generated within the MSC/NASTRAN environment. Four methods are available to generate SBV's. They are (a) Manual Grid Variation method (MGV), (b) Direct Input of Shapes (DIS), (c) Geometric Boundary Shapes (GBS), and (d) Analytic Boundary Shapes (ABS).

The shape optimization of adherends in bonded joints was attempted by employing the ABS method for the generation of SBV's. The design objective was chosen to be the minimisation of the Tsai-Hill factor, which is a general form of the Von Mises yield hypothesis. Yielding is said to occur when the *TSAI* value is greater or equal to 1. The reason for choosing this objective function was the goal of correlating the optimum shape obtained from the shape basis vector approach to that obtained by Groth and Nordlund (1991) using the optimization code OASIS-ALADDIN.

$$TSAI = \left(\left(\frac{\sigma_1}{X_t} \right)^2 + \left(\frac{\sigma_2}{Y_t} \right)^2 - \frac{1}{RHO} \frac{\sigma_1 \sigma_2}{X_t Y_t} + \left(\frac{\tau}{S} \right)^2 \right)^{\frac{1}{2}} \tag{5.22}$$

This function allows different weights to be given to stress component depending on their influence. X_t, Y_t and S are the tensile strength in the principal direction 1 and 2 and the in-plane shear strength respectively. That is, values of X_t, Y_t, S and *RHO* can be given different values to make the objective function being dominated for example by the shear stress. *RHO* is a non-dimensional factor. When this factor equals 1, the coupling of the direct and peel stresses has some weight on the *TSAI* value. On the other hand, when *RHO* is assigned a large magnitude, the effect of the coupling between the direct and peel stresses on the *TSAI* value is negligible. For the single lap joint analysis, the following yield stresses are used for the adhesive layer: X_t=Y_t=2 MPa, S=1.0 MPa and *RHO*=10000.

Details regarding the finite element analysis, material properties and loading are given by Rispler (1998). For the analysis of single lap joint failures, both peel and shear stresses are significantly important. The starting geometry was chosen to be square ended adherends so as to be able to obtain an equivalent comparison with the analysis run employing the heuristic approach ESO. Figure 5.4 depicts the finite element model mesh corresponding to the starting model. A coarse mesh was employed in order to minimize mesh distortion problems during the optimization runs, the figure shows thin elements representing the adhesive. The shape basis vectors were generated using an auxiliary boundary model. This model was generated by creating a group of bar elements on the boundary of the upper adherend along the length of the adhesive overlap. The auxiliary model was then loaded independently by placing a force at the end of the group of bar elements with boundary conditions corresponding to a cantilever beam. The design variables (node coordinates) along the adhesive overlap corresponding to the bottom adherend were related to their anti-symmetric design variables of the upper adherend to reduce computational effort.

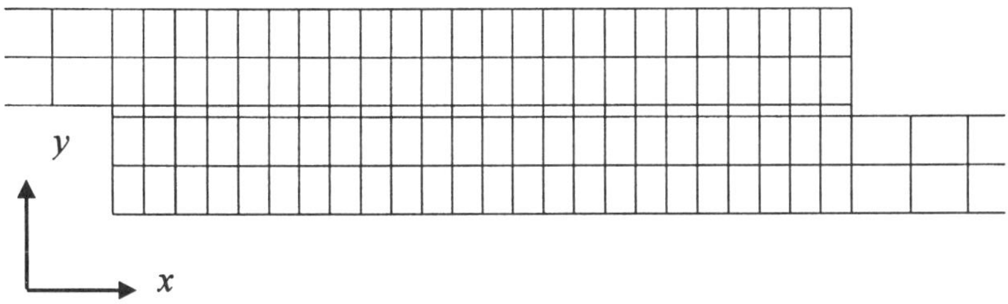

Figure 5.4 Mesh of starting finite element model

Initially the optimization was performed using four node elements. The result after the first design iteration was not encouraging. The finite element analysis could not be solved after the design variables (node coordinates of the upper and lower adherend along the adhesive overlap) were modified. The new coordinates assigned to the design variables resulted in an extremely distorted mesh with skewed and tapered elements well beyond the default allowable for these parameters. Essentially, the elements; in particular those close to the end of the overlap; had severe shape distortion resulting in ill conditioning of the stiffness matrix. In an attempt to restrict the mesh distortion, the auxiliary model which determines the SBV's was modified by generating different loads together with boundary conditions corresponding to simply supported beams. The type of element was also changed to eight node elements and the mesh refined locally at the tip of the adherends.

Unfortunately, the mesh distortion was only marginally reduced. It was then decided to modify the starting design by generating a new model with a single taper on both upper and lower adherend. The mesh corresponding to the single taper model is shown in Figure 5.5. A similar approach was followed by Groth and Nordlund (1991), where it was found that the design variables corresponding to the adherends; along the overlap length; needed to be assigned explicit linear constraints to ensure that the finite element mesh did not fail during the optimization process. Although the pre-profiling of the adherends removed all problems related to mesh distortion, no significant gain in stress reduction was achieved. Furthermore, the shape of the adherends did not vary apart from small coordinate variations close to the tip of the adhesive. That is, the optimum design converged to a single taper adherend, which is the known result.

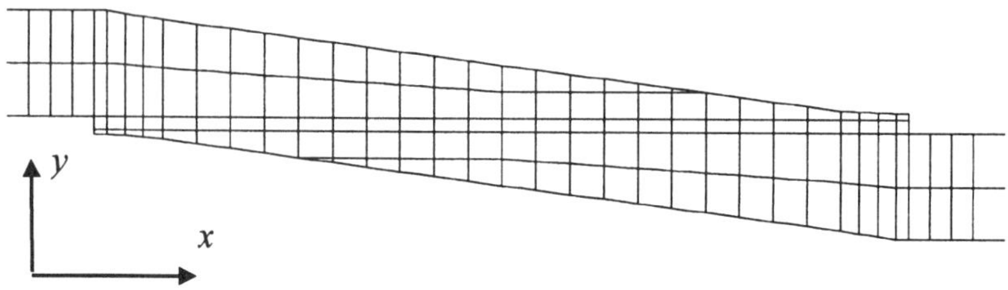

Figure 5.5 Mesh of finite element model for single lap joint

The results obtained from the shape basis vector approach show that this techniquc may only appropriate for fine-tuning existing designs and that radical shape changes can not be expected without the use of an automatic mesh generator. The regeneration of a model by fully re-meshing the structure once the optimization code detects severe mesh distortion could solve problems similar to the one attempted in this section.

5.6 APPLICATION OF THE ESO METHOD

The simple Evolutionary Structural Optimization (ESO) procedure developed by Xie and Steven (1997) has been employed for the optimization of adhesively bonded joints by shaping the adherend profiles to reduce peak stresses at the overlap ends. In the ESO procedure, a finite element analysis of the desired structure is performed and the stress distribution found. An element stress file is then set-up and using some criterion for rejection, here called a Rejection Criterion (RC), such as the Von Mises stress, the low stressed material is removed. For example, elements that have a Von Mises stress lower than a Rejection Ratio (RR) times the maximum Von Mises stress in the structure are deleted. The finite element analysis and rejection cycle are then repeated until a steady state (SS) is reached. An Evolution Rate (ER) is then introduced and added to the current RR until another steady state is reached. This process is repeated until the desired optimum is reached, for example, till all stress levels are 20 % of the maximum in the whole structure. The optimization software EVOLVE was used to optimize single and double lap joints as well as adhesive spew fillet as given in detail in Rispler (1998). In the following, only brief numerical results will be presented.

5.6.1 Shape optimization of adherends in single and double lap joints

The evolutionary structural optimization method described in the preceding section was applied to both single and double lap joints (Rispler, 1998). As mentioned previously, this method consists essentially in deleting the lowly stressed elements from the joint progressively. This should lead to fully stressed adherends which in turn re-distribute the stresses to the adhesive interface more evenly thus minimising the peak peel stresses normally found in this type of joints. In the cases presented, the overlap length and adhesive thickness were set as constants and only the adherends along the overlap were allowed to evolve. A single element across the adhesive layer was employed.

(a) Single lap joint

The dimensions of the joint together with the loading and kinematic constraints are shown in Figure 5.6. A plane strain analysis with 4-node quadrilateral (QUAD4) elements was employed. The load was corresponded to a nominal shear stress in the adhesive of 1 MPa and the materials had the following properties: aluminum adherends (E=70000 MPa, ν=0.3), epoxy adhesive (E=2000 MPa, ν=0.3). The geometrical non-linearity present in single lap joints was not accounted for in the analysis.

The starting models had square ended adherends with only one element across the adhesive thickness. A coarse mesh was initially employed to determine whether the size of the elements would affect the overall shape of the adherends. Furthermore, the finer mesh provides more points at the edge of the adherends for curve-fitting the profile.

The evolution criteria employed for the optimization of adherend shapes in both single and double lap joints was the Von Mises stress. The software allows the nomination of non-design properties for areas of the joint for which no elements are allowed to be deleted. The only area from which material could be removed was above/below the adhesive overlap length. Furthermore, in the evolutionary algorithm an additional check is undertaken to ensure that elements are only removed from the surface. This is called nibbling and ensures that no holes are created, it was demonstrated by Xie and Steven (1997) that the ESO nibbling algorithm is equivalent to finding the surface shape that minimizes the maximum surface stress which in turn makes the surface an iso-stress line which is the optimum shape.

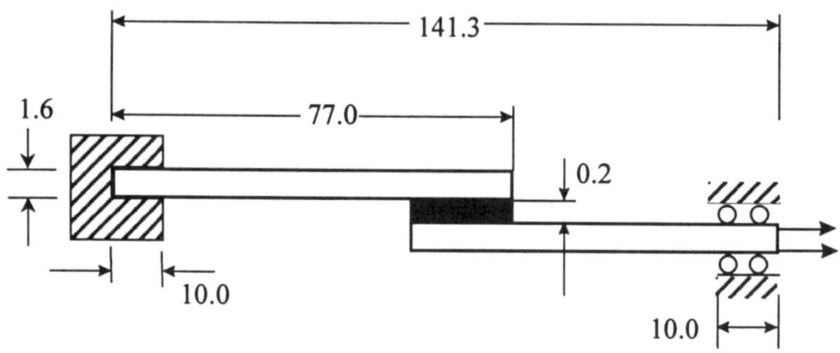

Figure 5.6 Geometry and boundary conditions of SLJ.

The optimum adherend shape for the single lap joint using the ESO is shown in Figure 5.7. The optimum rejection ratio was found to be RR=13.1%, yielding a reduction in peel stress at the end of the overlap of approximately 36%. The evolutionary rate was set to ER=0.1.The maximum shear stress increased by only 2% and its peak was shifted inwards to the joint by one element (0.2mm). For the purpose of comparison, the optimum shape obtained by Groth and Nordlund (1991) is redrawn and schematically depicted in Figure 5.8. It is clear that the present optimum adherend profile in Figure 5.7 is very similar to that obtained by Groth and Nordlund (1991) in Figure 5.8.

In order to obtain a more defined shape of the adherend profiles and to assess the influence of size of element on the optimum adherend shape, a refined mesh model was run. The refined mesh model was created by un-evolving the optimum solution obtained from the coarse mesh model by 10 steady states and subdividing each element along the overlap region in 9 elements (3x3). This model was employed as the starting geometry for the optimization run. The optimum adherend shape resulting from running the refined mesh model is shown in Figure 5.9. Although both optimum adherend shapes obtained from the refined mesh model and the coarse mesh model are very similar, the former illustrates more clearly the convexity of the adherend at the ends of the overlap. It can be noticed that within the overlap region the combined thickness of the adherends is equal to that of a single adherend outside the overlap region.

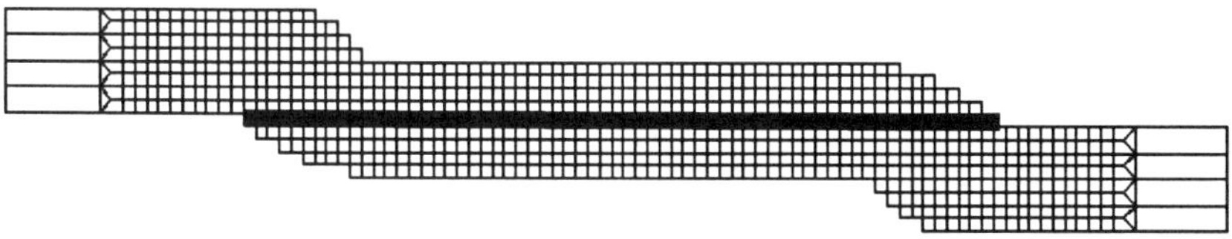

Figure 5.7 Optimum adherend profiling along overlap obtained using the ESO

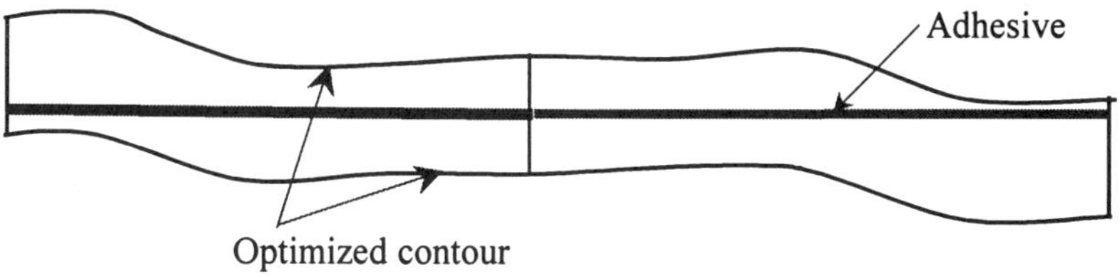

Figure 5.8 Optimum adherend profiling along overlap (redraw from Groth and Nordlund, 1991).

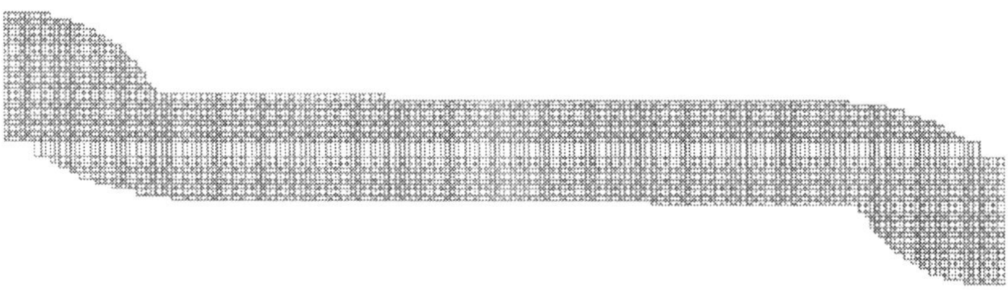

Figure 5.9 Optimum adherend profile for single lap joint with refined mesh

Figure 5.10 shows the shear and peel stress distributions versus the overlap length for both original and optimized model with a coarse mesh. Whilst it could be said that there is not a large reduction in the maximum shear stresses there has been a 40% reduction in the peel stress; at the same time the mass of material involved in the adherends has been reduced by as much as 50%.

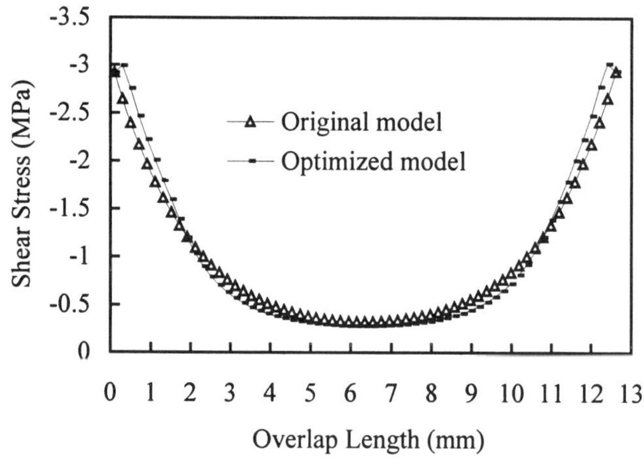

(a)

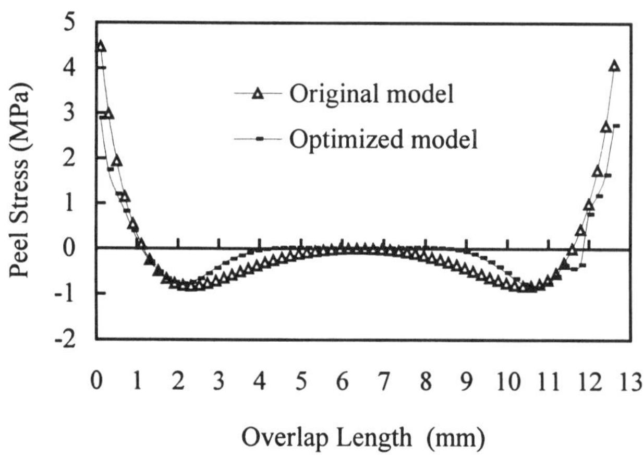

(b)

Figure 5.10 Shear and peel stress distribution for the original and optimized single lap joint

(b) Double lap joint

The double lap joint model was similar to the single lap joint one, but with different boundary conditions as shown in Figure 5.11. In this case, only the upper adherend was allowed to evolve so as to keep boundary conditions due to symmetry. The optimum adherend shape is shown in Figure 5.12.

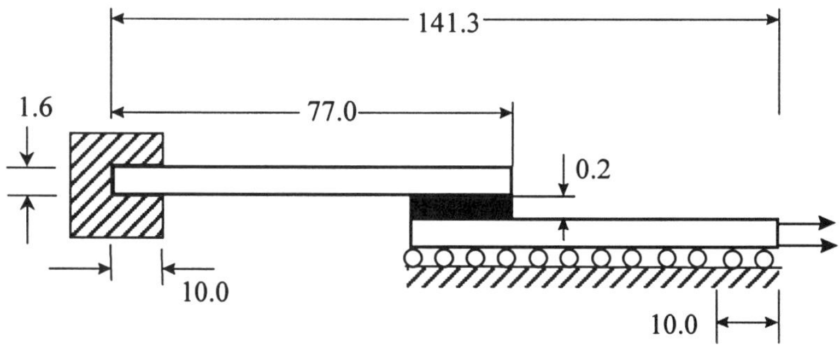

Figure 5.11 Boundary conditions for the double lap joint

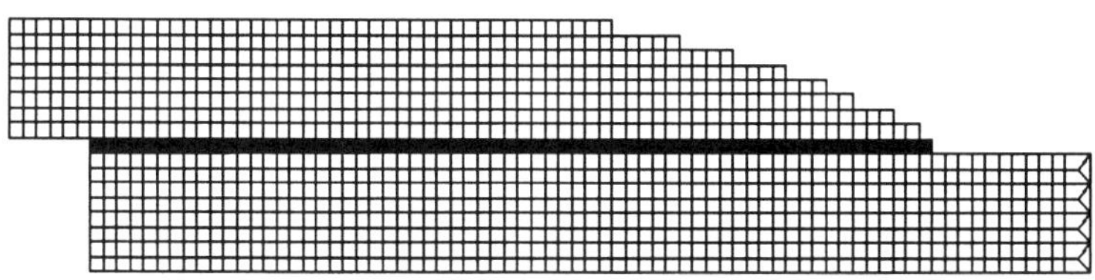

Figure 5.12 Optimum adherend profiling along overlap for the double lap joint

For the double lap joint, a reduction in peel and shear stress of 17 % and 8 % respectively was achieved with a RR=16.4 %. Figure 5.13 plots the shear and peel stresses model versus overlap length for the original and the optimized. The sharp drop in shear stress magnitude shown in Figure 5.13(a) at the tip of the adhesive occurs as a result of the adherend element above the adhesive tip being removed.

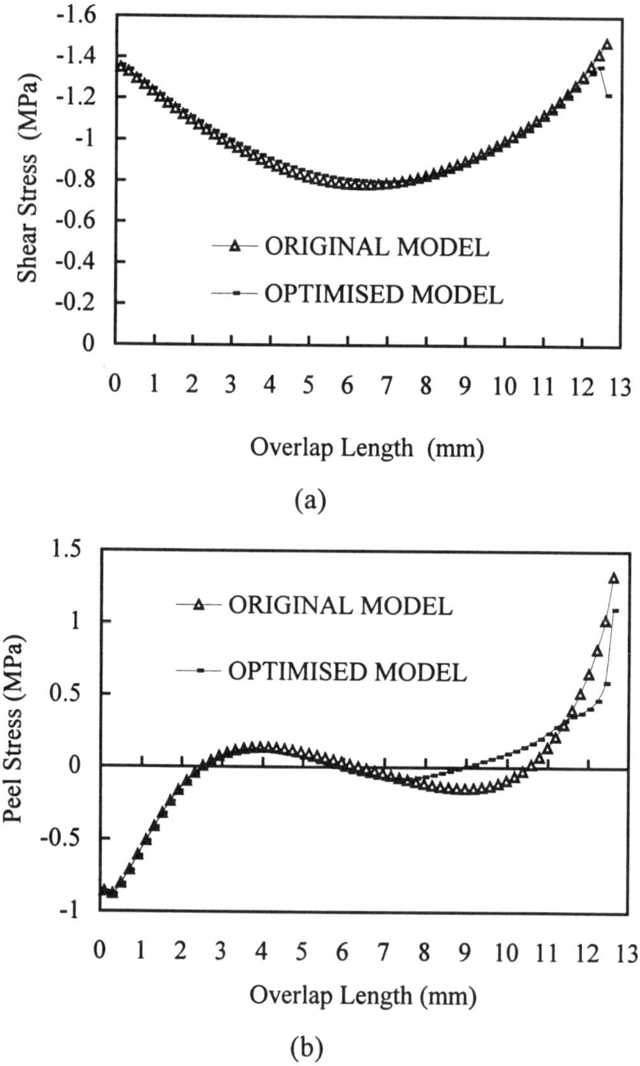

Figure 5.13 Shear and peel stress distribution for original and optimized double lap joint

Similarly, the optimum upper adherend shape for the double lap joint analysis was un-evolved and re-analyzed. The optimum adherend shape resulting from running the refined mesh model shows that there exists a larger extent of the convex curvature on the optimized upper adherend when comparing it to the coarse mesh model. While the shape of the upper adherend remains straight up to two thirds of the overlap length on the coarse model, on the refined model the flat region only extends up to half the overlap length. This results in a less steeper taper for the refined mesh model.

Most work on optimization of bonded joints has tried to achieve a uniform shear stress distribution along the overlap. In practice, a uniform shear stress distribution is not desirable due to fatigue and creep considerations (Hart-Smith, 1987). In single lap joints, the most critical stresses are the peel stresses, which are induced by the eccentricity of the load path. Similarly, the double lap joint also suffers from peel stress concentrations, but in this case the peak shear stresses are also important. The above results indicate that the methodology was successful in reducing critical stresses in both the single and double lap joint configurations. The achieved reductions in peak peel stresses in the single lap joint and shear and peel stresses in the double lap joint should ultimately lead to an increase in joint strength by reducing the stress concentrations causing failure initiation.

5.6.2 Shape optimization of adhesive fillets

Adams and Peppiatt (1974) investigated the effect of spew fillet size on stresses by assuming a 45° fillet and changing its size. The magnitude of principal stresses obtained from such analysis showed a large difference when compared to results obtained from analysis which assume the adhesive to have a square edge. The highest principal stress was found to lie adjacent to the tip of the unloaded adherend (upper adherend). This correlated well with observed failure characteristics during testing. The inclusion of a triangular spew fillet (45°) decreased the magnitude of the maximum principal stress by 40% when compared to a square end adhesive fillet. Crocombe and Adams (1981) included the effect of additional parameters such as material and geometric properties. In their analysis, the stress distribution across the adhesive thickness was not assumed constant. Their results for a lap joint with a spew fillet indicate that the averaged maximum adhesive stress increased with decreasing modulus ratio (increasing adhesive stiffness) up to a limiting value, after which they decreased again. Variations of overlap length, adhesive thickness and adherend thickness was also included in their analysis. The effect of spew fillet on stresses in adhesive has also been discussed in Chapter 2. In this section, we will use the ESO to optimize the size of the spew fillet.

Let us consider the joint with its geometry of the adherends and adhesive shown in Figure 5.14, and the goal is to minimize the maximum principal stresses in the adhesive fillet found in tabs of tensile test specimens with reverse taper by applying the ESO method. This should ultimately increase the joint strength. The central adherend is carbon fiber reinforced plastics (CFRP) and two types of materials are used for the upper adherend, namely, the same CFRP and titanium. In addition, the Young's modulus for the adhesive is varied to study the effect of adhesive property. In all cases, the starting model corresponded to a square fillet profile. The mesh employed for all different cases was exactly the same. Some distorted elements near the upper adherend tip were present due to an extremely fine taper. Due to symmetry, boundary conditions same as that for the double lap joint were employed and shown in Figure 5.15. The loading applied to the model was 1 GPa in tension and 100°C cool-down on all nodes. The cool-down temperature aims at including the effect of residual thermal stresses due to cure shrinkage.

Figure 5.14 Adherend and adhesive geometry

The material used for the central adherend is unidirectional prepreg T800 with the following properties: E_1=175 GPa, E_2=E_3=10 GPa, ν_{12}=0.27, ν_{13}=ν_{23}=0.0154, $\alpha 1$=-0.9 $\mu\varepsilon/°C$, $\alpha 2$=$\alpha 3$=27 $\mu\varepsilon/°C$, X_1=2.7 GPa, and X_2=X_3=50 MPa (α represents the coefficient of thermal expansion). The upper adherend is taken to be titanium with the following material properties: E=110 GPa, ν=0.29, α=9 με/°C and the maximum Von Mises stress allowable = 1000 MPa. The adhesive has the following properties:

E=3.8 GPa, ν=0.38, α=60 $\mu\varepsilon/°C$ and the maximum tensile and compressive stresses = 40 and 52 MPa respectively.

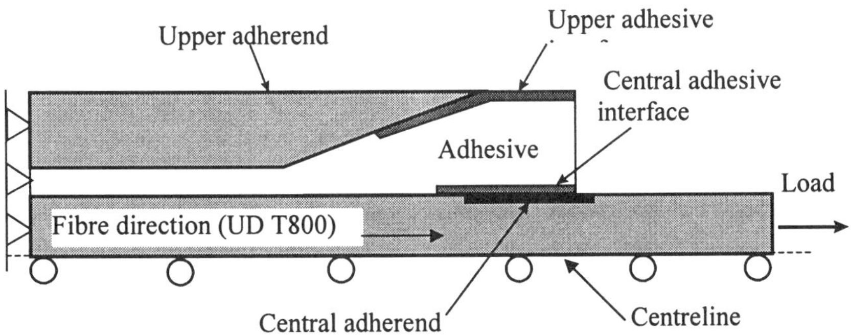

Figure 5.15 Schematic of monitoring properties for starting model (steady state=0)

The EVOLVE code was employed to optimize the shape of the adhesive fillet. The driving criterion for all models was the mean maximum principal stress in the structure. The only material which was allowed to evolve was the adhesive. The adherends were selected as non-design properties. The evolutionary rate was selected to be 0.1 %. This was found to be necessary as with higher evolutionary rates, too many elements would be deleted from the structure between steady states.

Within the two cases analyzed (different upper adherend materials), two different optimization techniques were employed. The first one consisted in allowing the optimization code to consider the stresses of all elements in the model to derive the driving criterion (Average Maximum Principal Stress). This technique will be referred to in this Section as the Classical Evolutionary Technique. The second technique consisted of employing the passive element capability of EVOLVE. The passive element approach allows for the selection of desired properties from which the stresses are calculated. Essentially, as the objective is to minimize stresses in the adhesive, the adherend properties are selected as passive so that the stresses calculated for elements within the adherends do not influence the driving criterion. This approach also simplifies the post-processing significantly. This technique will be referred to in this Section as the Passive Evolutionary Technique.

In the Classical Evolutionary Technique, three properties were selected to monitor the maximum principal stresses during the optimization process. These properties are shown in a schematic (Figure 5.15) of the starting model (square ended adhesive fillet shape, steady state = 0) and were selected in areas where maximum joint stresses were expected. The dark grey colour represents monitoring properties corresponding to the adhesive while the remaining monitoring property (black shaded) corresponds to the central adherend. The EVOLVE code then allows us to look at every steady state for the maximum principal stresses recorded in the adhesive and in the adherend. This reduces the post-processing of the results quite considerably and also makes the selection of the optimum fillet shape a simpler task. Although the maximum principal stresses for the initial structure (SS =0) were found to be either on the adhesive in contact with the upper or lower adherend, the position of this stress varies in some steady states. This means that the maximum principal stress might not occur in the monitoring properties selected. In these cases, a full finite element analysis has to be run again for the chosen steady states. The selection of all adhesive plates as monitoring properties would prove prohibitive due to the required memory to store all stress values for all steady states.

As the stresses of the adherend elements were not included in the optimization routine, the Passive Evolutionary Technique did not require the use of monitoring properties. The results file from the optimization process allows for determination of a global optimum by plotting a single graph of the maximum and mean values of the driving criterion for all steady states. This is only applicable when the variable to be minimized (optimized) coincides with the driving criterion.

The results obtained from applying the Classical Evolutionary Technique showed some variation between the optimum shapes arrived at by using different mesh densities. Instead, the Passive Evolutionary Technique was very consistent for both the course mesh and fine mesh models analyzed. Furthermore, it was found that the selection of global optimum was greatly simplified by employing the Passive Evolutionary Technique. The selection of a global optimum by employing the latter technique consisted of simply producing a plot of the maximum principal stress versus each steady state of the optimization process, straight from the results file. Therefore, the results presented later correspond to analysis carried out using the Passive Evolutionary Technique.

The aspect ratio of the elements has to be kept as constant as possible so that the effect of this variable on the fillet shape is minimized. The size of the elements greatly influences the time taken to obtain an optimum solution. Therefore a compromise has to be made between element size, solution time and accuracy of proposed optimum shape and stress magnitudes. An alternative to obtain the final shape of the optimum fillet is to run a B-spline through the nodes of elements at the edge of the adhesive fillet. Again, the accuracy of this spline will depend on the mesh density.

The Passive Evolutionary Technique allows for selected properties to be chosen to drive the optimization process. All the adherend elements were assigned passive properties so that the stresses calculated for these elements had no effect on the driving criterion employed during the optimization process. The mean principal stress was employed as driving criterion for all models analyzed. The passive element approach offers a great advantage in terms of post-processing the results. As the stresses of the adherend elements are not included in the optimization routine, the results file from the optimization process allows for determination of a global optimum by plotting a single graph of the maximum and mean values of the driving criterion for all steady states. This is only applicable to analysis where the driving criterion is the variable to be minimized. In the present case the objective is to minimize the maximum principal stress in the adhesive fillet. The optimization process calculates the maximum and mean stresses; chosen as driving criterion; for all steady states (SS) and writes these results to a file.

(a) CFRP upper adherend, CFRP lower adherend and Young's Modulus of adhesive E=3.8 GPa

The starting model has a square ended adhesive fillet. The maximum and mean principal stresses were plotted from the result file of the optimization program yielding the steady state with the minimum value of the maximum principal stress (global optimum).

Figure 5.16 depicts both the maximum and the mean principal stresses during the complete optimization process pertaining to the coarse mesh model. The graph indicates that up to SS 80 not much stress reduction is achieved. This is due to the large overlap distance employed in the baseline model. As the overlap length decreases, the maximum principal stress decreases as well. A small plateau is reached at approximately SS 160 where the adhesive fillet is close to its global optimum. Finally, the global optimum is reached at SS 167. The rejection ratio at this SS is 94.9% whereas the maximum principal stress in the adhesive is equal to 53.98 MPa. A reduction of approximately 55 % in maximum principal stress is achieved; when comparing to the baseline; by this method yielding a 45° fillet as optimum shape. The 45° fillet starts 1 element to the right of the end of the upper adherend tip/adhesive interface. The maximum principal stress in the adhesive fillet is found halfway through the fillet and also at the interface with the lower adherend. The shape of the fillet together with the stressing of the fillet for the optimum configuration is shown in Figure 5.17.

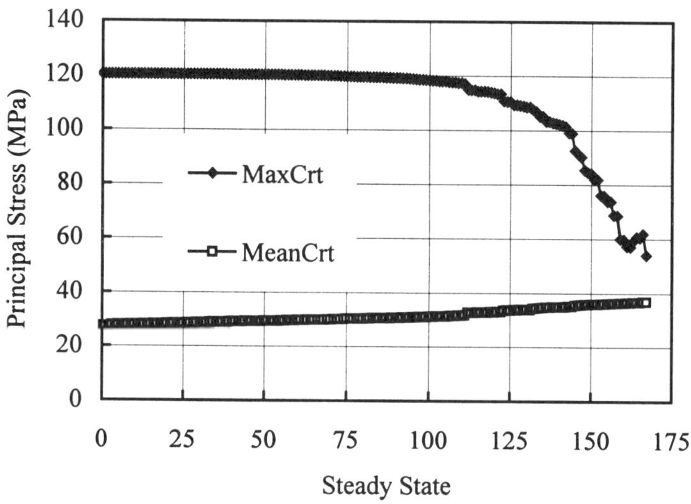

Figure 5.16 Maximum and mean principal stresses at adhesive fillet obtained using a coarse mesh

The peel stresses at the adhesive and CFRP lower adherend were also compared to the baseline values. It was found that the maximum peel stress in the adhesive was reduced from 42.69 MPa to 21.14 MPa (50 %) whereas the maximum peel stress in the carbon fiber reinforced adherend was reduced from 30.31 MPa to 18.26 MPa (40 %).

To assess the influence of the mesh size on the optimum shape of fillet and to obtain a more accurate representation of plate stresses, a finer mesh model was run. This model was created by un-evolving the optimum solution of the coarse mesh model by 3 steady states and re-meshing locally the adhesive fillet by sub-dividing every QUAD4 element into 9 QUAD4 elements. The same graphs were plotted to obtain the global optimum. The graph of the maximum and mean principal stresses corresponding to the fine mesh model is shown in Figure 5.18.

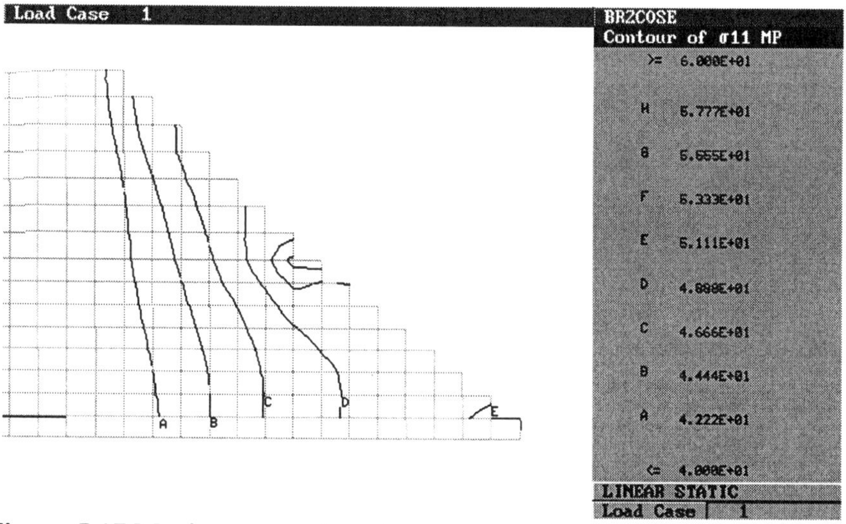

Figure 5.17 Maximum principal stress contours for optimum shape fillet at adhesive elements obtained using a coarse mesh

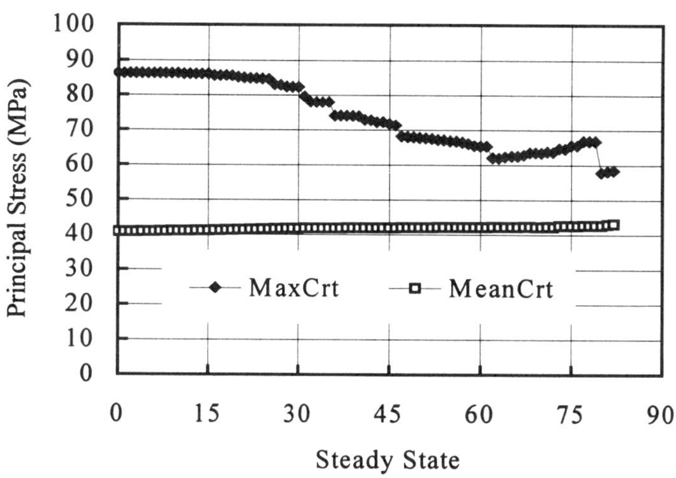

Figure 5.18 Maximum and mean principal stresses at adhesive fillet obtained using a fine mesh

Despite an expected increase in the magnitudes of maximum principal stress on the adhesive fillet from 53.98 MPa to 57.79 MPa, the fillet shape remains the same as for the coarse model. The maximum peel stress in adhesive has a slight increase from 21.14 MPa to 22.59 MPa while the maximum peel stress in the CFRP adherend decreases slightly from 18.26 MPa to 17.41 MPa. The shape of the optimum fillet with the contours of maximum principal stress on the adhesive fillet is depicted in Figure 5.19. It is clear that there is no significant change in the shape and the contours of maximum principal stress.

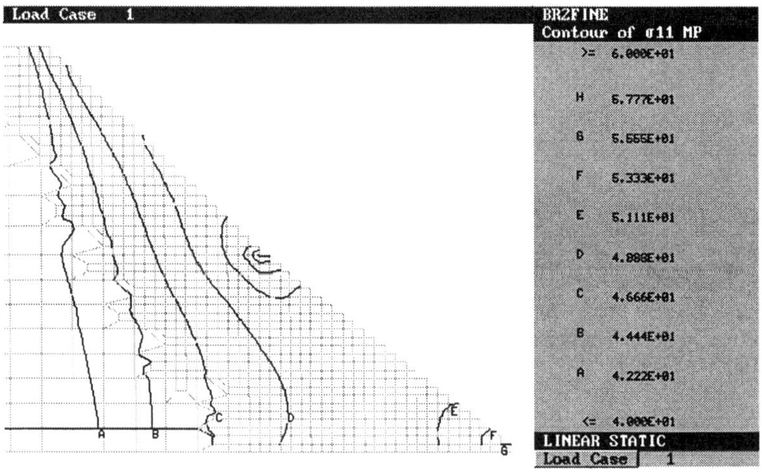

Figure 5.19 Maximum principal stress contours for optimum shape fillet at adhesive elements obtained using a fine mesh

(b) Titanium upper adherend, CFRP lower adherend and Young's Modulus of adhesive E=3.8 GPa

This case aims at assessing the influence of only replacing the CFRP upper adherend with a Titanium one. The fine mesh model was created by unevolving the optimum solution of the coarse mesh model by 3 steady states and re-meshing locally the adhesive fillet by sub-dividing every QUAD4 element into 9 QUAD4 elements. Figure 5.20 plots the maximum and mean principal stresses, which indicate that changing the upper adherends from CFRP to titanium results in a decrease in maximum principal stress from 57.79 MPa to 52.4 MPa (10 %). Figure 5.21 depicts the contour of the maximum principal stresses and the optimum fillet shape. It is seen that the effect of changing the adherend material from CFRP to titanium reduces the tapering angle of the spew fillet from 45° to 30° approximately, with a slightly lower angle near the central adherend interface.

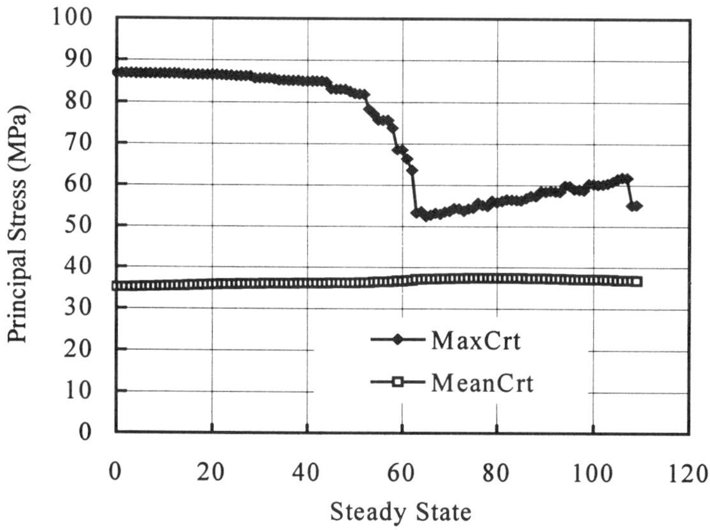

Figure 5.20 Maximum and mean principal stresses at adhesive fillet obtained using a fine mesh.

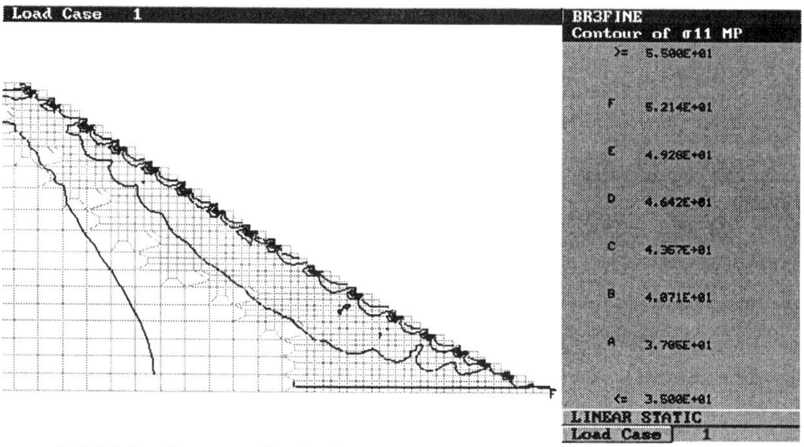

Figure 5.21 Maximum principal stress contours for optimum shape fillet at adhesive elements obtained using a fine mesh

(c) Effect of adhesive Young's Modulus

To study the effect of adhesive modulus on the optimum solution, the Young's modulus of the adhesive is varied. Optimum solutions are obtained using the fine mesh for the case of titanium upper adherend with adhesive having a Young's modulus of 2.6 and 5.0 GPa, respectively. Table 5.1 compares the tapering angles of the spew fillet, the maximum principal stress in adhesive, the maximum peel stress in adhesive, and the maximum stress in the CFRP central adherend. It is seen that for the cases studied here the optimum tapering angle is between 30° and 45°. Comparing the results of the last three rows reveals that the maximum principal stress in adhesive becomes smaller when the adhesive becomes softer, while the maxim peel stresses in adhesive and the central adherend becomes slight larger. It is also clearly shown that reductions in maximum principal stresses and in peel stresses in both the adhesive and the central CFRP adherend are obtained by replacing the CFRP upper adherend with a titanium one. This is believed to be due to the increase in stiffness in the through thickness direction. This increase in stiffness helps reduce the deformation in the *y* direction, thus causing a reduction in peel stresses.

Table 5.1 Comparison of solutions of the baseline and optimized models

Upper adherend material	Young's modulus adhesive (GPa)	Max principal stress in adhesive (MPa)	Max peel stress in adhesive (MPa)	Max peel stress in central adherend (MPa)	Tapering angle of spew fillet (Deg)
CFRP[1]	3.8	120	42.69	30.31	/
Titanium[1]	3.8	98.90	42.60	30.25	/
CFRP[2]	3.8	57.79	22.59	17.41	45
Titanium[2]	2.6	43.26	16.97	13.05	45
Titanium[2]	3.8	52.40	16.28	10.64	30
Titanium[2]	5.0	61.78	14.68	10.14	30

[1] Baseline model; [2] optimized model.

6

Effect of transverse stitching

6.1 INTRODUCTION

The high positive peel (normal) stress near an overlap end and the weak interlaminar strength are believed to be the two major contributing factors to the initiation and propagation of delamination along the interface between the surface and the second ply in a composite adherend in bonded composite lap joints. Figure 6.1 schematically depicts the onset and propagation of interlaminar delamination between the surface and second plies. Depending on the joint configuration and loading conditions, the delamination can grow along the interface or even jump into the next interface, or the deformed surface ply can fracture. It has been shown in Chapter 4 that the presence of small delamination crack can result in a reduction of the failure load for lap joints. From this origin, the onset of interlaminar delamination starts the failure of the lap joints and the corresponding loads are the failure loads. Thus the joint strength for some joint configurations can be limited by the interlaminar strength, which is the weak link for composite adherends as it relies on the brittle matrix tensile properties and the bonding strength of the fiber/matrix interface. To achieve increased joint strength, it is necessary to choose an optimum combination of parameters in material, design and manufacturing to reduce the magnitude of the positive peel stress at the overlap ends and to increase the interlaminar strengths.

Interlaminar strength of laminated composite materials can be improved by increasing the fracture toughness of the composite laminate, which has led to development of toughened resins and interfacial coatings to enhance the interfacial strength, and to the hybridization of fibers in composite laminates (Dransfield et al, 1994).

As described in Chapter 5, tapering adherend near an overlap end is simple and effective in terms of reducing the peak positive peel stress and the peak shear stress. Adherend tapering can come in two forms, i.e., external and internal tapering. External tapering reduces the thickness of the adherend near the ends of the overlap, while internal tapering reduces the adherend thickness and increases the thickness of adhesive. For metal adherends, tapering can be achieved via machining process, while for a composite adherend, tapering can be realized by dropping plies gradually near the overlap end. Although end tapering is simple and effective, it can be labor intensive and less cost-effective, especially for composite adherend.

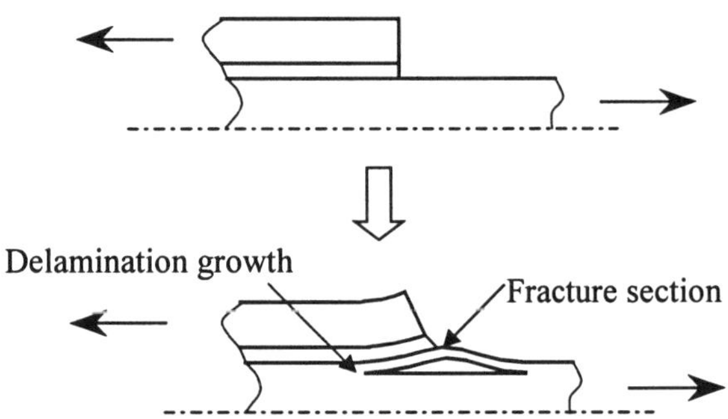

Figure 6.1 Peel stress induced interlaminar delamination in composite lap joints

Another straightforward concept is to simply apply compressive forces right at the overlap edges to counteract the positive peel stress. Sawyer (1985) designed an experimental set-up as schematically shown in Figure 6.2a and demonstrated that the application of transverse compressive forces at the overlap edges can significantly increase the joint strength. For example, Sawyer (1985) showed that when the total transverse compressive normal force is 25% of the failure load of a bonded titanium single lap joint with zero compressive normal force, the failure load of the joint actually doubled. Figure 6.2b demonstrates that delamination takes place outside the joint overlap due to the applied compressive normal forces at the overlap ends for a composite single lap joint.

While this idea is appealing, it is impractical for most bonded joints having metal adherends because of the external nature of the application of the transverse force. Composite joints can be internally transversely restrained using a stitching technique. Through-thickness stitches form loops and thus provide clamping-type forces or compressive forces when stretched.

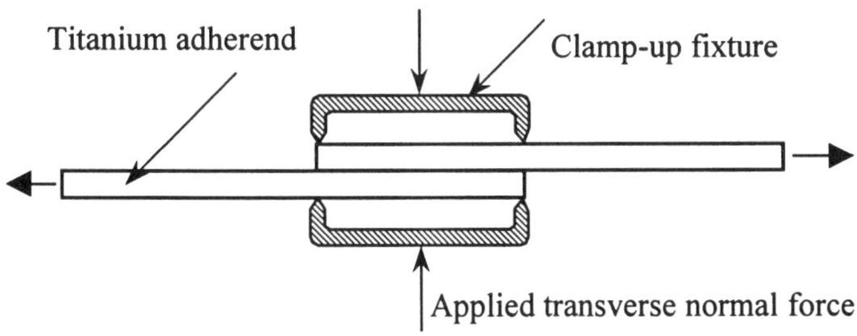

Figure 6.2a Joint set-up for applying transverse normal compressive forces (Sawyer, 1985)

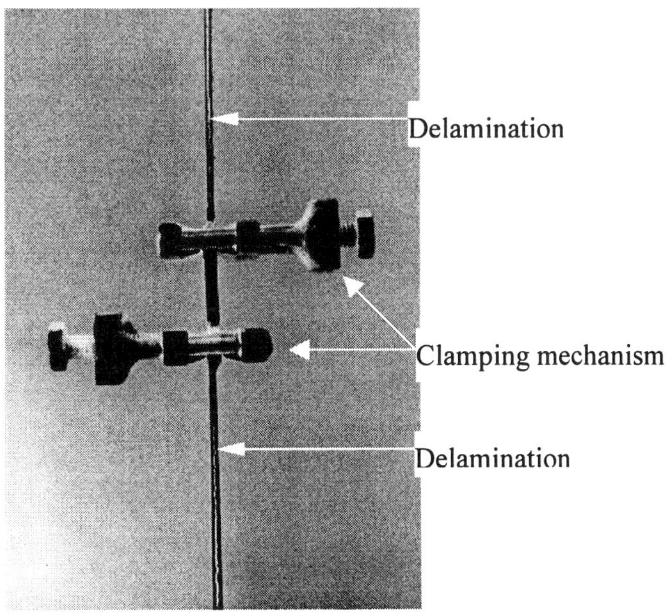

Figure 6.2b Delamination occurred outside joint overlap due to applied normal compressive forces

Joining of composite materials is a critical issue in designing and manufacturing of composite structures. Composite laminates or splices can be joined together using either secondary bonding or co-curing process. In the secondary bonding process, a layer of adhesive is applied, while in co-curing process two parts are bonded together simply by adhesion of resin or matrix between two adjacent surface plies. Similar to co-curing process, liquid molding is another manufacturing process. In liquid molding process, an integral composite part is prepared as a dry fiber preform and then placed in an enclosed mould for resin injection and consolidation. In both co-curing and liquid molding processes, adhesion of resin is the joining mechanism between two components or surfaces. By placing fibers between the two joining members via transverse stitching, we can make stitched joints or composites using either the co-curing or the liquid molding process.

In a strict sense, stitched joints are not adhesive bonded joints as there is no adhesive layer. However, as the load transferring media is the resin-rich epoxy layer, which may be regarded as equivalent to adhesive layer, stitched joints share many same characteristics as adhesive bonded joints. From this point of view, stitched joints maybe regarded as fiber or stitch reinforced bonded joints, or a mixture of bonding with epoxy and fastening with stitches. In a more general sense, stitched composites have the potential to significantly improve impact resistance (Mignery et al, 1985; Ogo, 1987; Liu, 1990; Morales, 1990; Drandfield et al, 1994), and to provide efficient, durable and affordable composite structures. It has been shown that transverse stitching can improve the static and fatigue strength of composite lap joints compared to the unstitched ones (Sawyer, 1985, Lee and Liu, 1990; Tong and Jain, 1995; Tong, 1998c; Tong, Jain et al, 1998). This chapter will discuss some fundamental issues related to manufacturing, testing and modeling of stitched composite joints.

6.2 MANUFACTURING

Composite laminates can be stitched in either prepreg state or in dry fiber preform. For the process of stitching prepreg, a number of prepreg layers are laminated and then transversely stitched using a sewing machine. Due to the high viscosity of the uncured resin in a prepreg state, the push-in and pull-out motion of the needle with thread can easily break most of the fibers within the diameter of the needle. Although stitching prepreg can improve fracture toughness of mode I significantly, it also leads to lower in-plane strengths (Ogo, 1987). In the process of stitching dry fiber preform, a stack of dry fiber layers is laminated to form a preform and then transversely stitched. The stitched preform is then placed in an enclosed mould for injecting liquid resin and consolidating the part. Because the preform does not contain sticky resin, fibers in the path of the stitching needle are readily pushed away from the stitch hole area when the needle is pushed into the preform. As will be shown later in this chapter, only a small amount of fibers are broken during the stitching process. Thus in stitching dry preform, damage is significantly reduced compared to stitching prepreg. It is thus desirable to stitch preform rather than to stitch prepreg. In this section, some of the main manufacturing factors related to stitched composites will be discussed. These factors include stitching pattern, stitching parameters, and liquid molding process.

6.2.1 Stitching pattern

There are two basic types of stitches, one is known as lock stitch and the other known as chain stitch (Dransfield et al 1994). For composite only the lock stitch is used. Lock stitch consists of a loop of two threads between the needle and the bobbin threads, which comes from both the top and the bottom of the stitched materials, as schematically shown in Figure 6.3(a).

For composite applications, positioning the intersection in the middle of the fabrics may create a resin rich and a high stress concentration point. One solution for overcoming this problem is to modify the lock stitch by forcing the needle thread loop to travel through the fabrics and to form the two-thread loop on the surface of the fabrics. This can be achieved by either tensioning the bobbin thread or easing the needle thread. A schematic diagram of a modified lock stitch is shown in Figure 6.3b. The modified lock stitch has frequently been used for composite applications.

There are four basic types of stitching patterns that are common with most domestic or industrial sewing machines as shown in Figure 6.4. These are straight plain stitching, zigzag stitching, box zigzag stitching, and plain zigzag stitching. In the straight plain stitching of Figure 6.4a, the pitch length is the major geometrical parameter, which defines the distance between two consecutive stitching holes. In the zigzag stitching shown in Figure 6.4b, the stitch threads travel continuously in a zigzag form between two parallel lines. There are three parameters, the pitch length, the stitch width and the stitch length. Similarly, the pitch length represents the distance between any two consecutive stitching holes. The stitch width defines the distance between the two parallel lines. The stitch length measuring the distance between every two subsequent stitching holes along either of the two parallel lines. It should be mentioned that only two of the three parameters are independent. The box zigzag stitching shown in Figure 6.4c is a modified zigzag stitching. In addition to the three parameters in zigzag stitching, one more parameter is required to describe the box zigzag stitching. Another variation of the zigzag stitching is shown in Figure 6.4d, where plain straight stitching can be applied between any two consecutive stitching points.

(a) Lock stitch

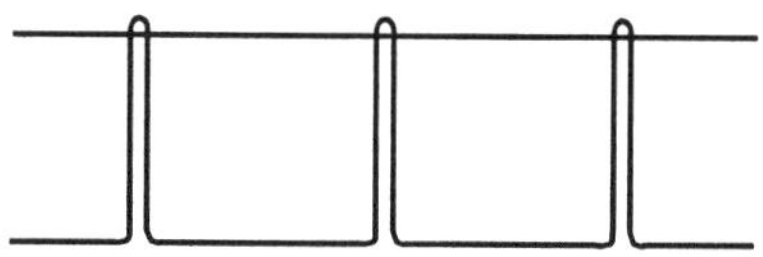

(b) Modified lock stitch

Figure 6.3 Schematic diagram of lock and modified lock stitches

Design of a stitching pattern for a structural part represents an optimal combination of various basic stitching patterns, such as these shown in Figure 6.4. For example, Figure 6.5 shows the top view of a plate stiffened with one stiffener and reinforced with a selected pattern of transverse stitching. As shown in the figure, a zigzag stitching pattern can be used along the edge of the stiffener flange and two rows of plain straight stitching may be used along the flange.

Stitching density is an important stitching parameter that defines the number of stitches per unit area. It is dependent of pitch length, stitch width and stitch length as well as stitch row spacing. It has been shown that increasing stitch density can lead to an increase in mode I fracture toughness (Ogo, 1987; Morales, 1990).

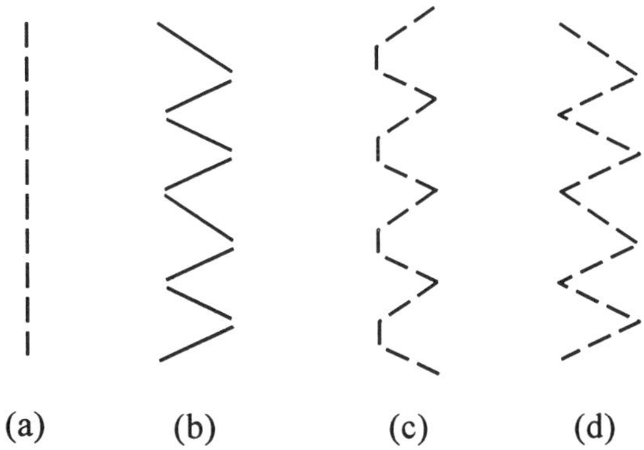

Figure 6.4 Four types of basic stitching patterns, (a) plain straight stitching, (b) zigzag stitching, (c) box zigzag stitching, and (d) plain zigzag stitching.

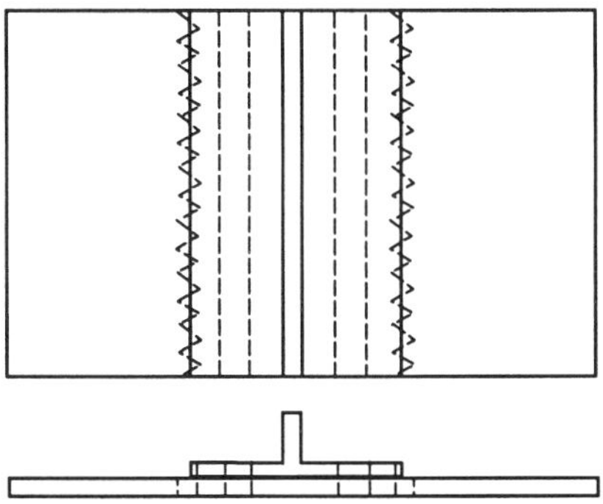

Figure 6.5 Top view of a stiffened and stitched plate

Conventional sewing machines only provide vertical transverse stitches; namely, all stitches are perpendicular to the mid-plane of adherends. A new concept is to place the stitch thread at an angle to the mid-plane, which is referred to as angle stitching. Perhaps one important application is to apply angle stitching where thickness discontinuities exist. Another motivation for angle stitching is to provide reinforcement for Mode I and Mode II type fracture as opposed to vertical stitching only providing reinforcement against Mode I deformation. Tong and Steven (1996) preformed a numerical study on the effect of stitch angle on the peak stresses in stitched composite joints.

6.2.2 Stitching parameters

Along with the design of stitching patterns, an engineer needs to select thread material and to select the sewing needle as well as the pretension in the stitch. When selecting thread material, it is important that the selected thread has sufficient amount of extensibility and abrasion resistance so that it can survive the bending deformations and stresses when passing through the eye of the needle. Glass, Kevlar and carbon fibers are commonly used. When a thread material is selected, the next step is to select the thread ply, thread diameter and thread finish. The thread ply is defined as the number of yarns twisted together to form a single stitching thread. Ogo (1987) and Morales (1990) have discussed the effect of these parameters.

Pretension in stitch is another important stitching parameter as it has been shown that pretension directly affect the compressive forces created by the stitch loops and has an important effect on reducing the peak peel stress in lap joints (Tong and Jain, 1995).

6.2.3 Resin transfer moulding

Resin transfer moulding (RTM) is one of the promising liquid moulding techniques that are being used to make composite structures in an efficient and cost-effective way. RTM is a closed mould process in which liquid resin is pumped into an enclosed mould containing the structural fiber preform. One of the major processing requirements for RTM is the processability of a resin system primarily described as its flow. A high degree of resin flow is required to make good fiber/matrix close contact and minimize voids. The two major contributing factors are the low resin viscosity and long pot-life at processing temperatures. Some of the advantages of RTM over conventional prepreg process include low labor cost, low raw material cost, superior productivity etc. RTM allows the fabrication of more complex structural shapes and achieves close dimensional tolerance.

6.3 SINGLE-LAP JOINTS – EXPERIMENTAL RESULTS

Two sets of experimental results will be presented in this section to show the effect of through-thickness stitching on strength of single-lap joints manufactured by stitching preform and prepreg (Tong et al, 1998; Sawyer,

1985). Both results indicate a promising effect of the through-thickness stitching in terms of improving joint static and fatigue strength, although the material and geometrical dimensions used in the two tests are different.

6.3.1 Joints fabricated by stitching preform

(a) Lap joints with thin adherends

The single-lap joint specimen is shown schematically in Figure 6.6. The specimen has an overlap of 30 mm and a width of 25.4 mm. The unsupported length is either 70 or 90 mm.

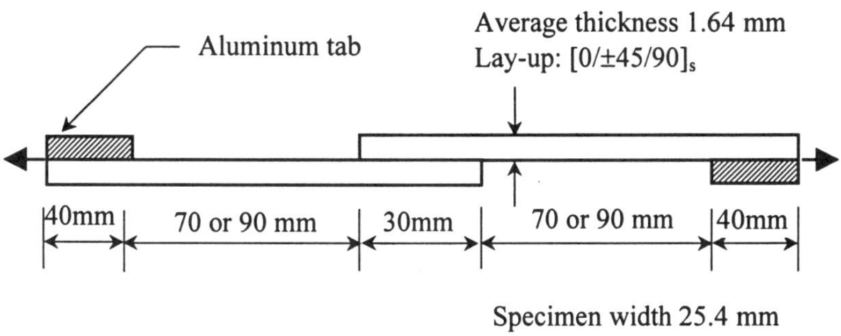

Figure 6.6 Geometrical configuration of the single-lap joint specimen manufactured using the RTM process

All specimens were manufactured from Ciba Composites Injectex® uniweave carbon fabric GU230-E01 and GY260 epoxy resin/HY917 hardener/DY070 accelerator. The uniweave material has 90% of its fibers oriented in the warp direction and the remaining fibers in the weft direction to hold the warp fibres in place for ease of handling. The Injectex® has been developed for precise fabric placement at preform stage prior to resin infusion. The stitch material used was a twisted 40tex (2×20) Kevlar™ thread, and zigzag stitching pattern was employed with the overstitch limited to 1 mm as schematically shown in Figure 6.7a. Figure 6.7b depicts the typical photograph of the failed stitched specimens.

In preparing the specimens, the following steps were used: (a) overlaying two $[0/\pm45/90]_s$ fabric stacks followed by debulking under vacuum and heat to produce a preform of single-lap panel; (b) applying transverse stitches following the designed pattern; and (c) injecting resin and consolidating the panel under clamping pressure and a curing temperature of 80°C for 4 hours. All panels were then post-cured, dried and C-scanned prior to being cut into specimens. Aluminum end tabs were bonded on one side of the adherend using an instant epoxy, as shown in Figure 6.6, to provide a co-linear loading path for the specimens.

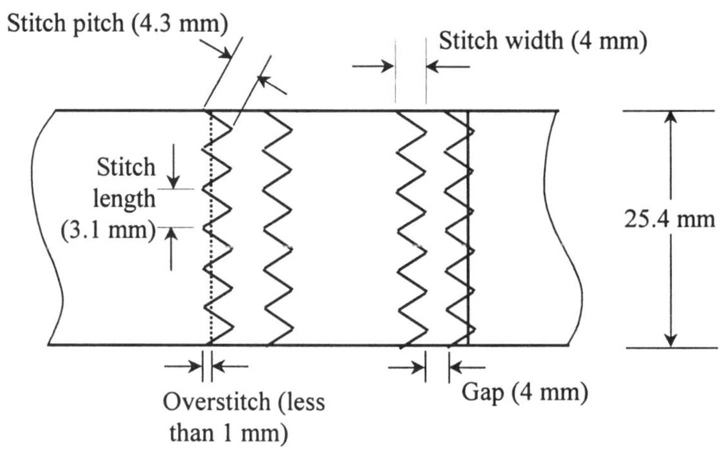

Figure 6.7a Top view of the four-row zigzag stitch used in the overlap of the single-lap joint

Figure 6.7b Top view of a stitched single-lap joint specimen failed under tensile load

Figure 6.7c Photograph of typical fracture surfaces of stitched specimens

Figure 6.7d Typical photograph of a broken stitch in stitched specimens

All specimens were tested in tension on an Instron testing machine at room temperature with a loading rate of 0.5 mm/min. The fractured surfaces were examined under scanning electron microscope (SEM) to identify the failure mechanisms for both unstitched and stitched specimens. Figure 6.7c shows a typical fracture surface of stitched joints while Figure 6.7d depicts a single broken stitch.

The measured axial loads increased almost linearly with the applied axial displacement for all specimens up to the final failure accompanied by a bang. For all specimens catastrophic failure occurs upon attaining the ultimate load. Delamination is observed to take place at the interface between the two adherends and breakage of stitching threads is also noticed. Missing fibers and matrix are also observed from both adherends by careful inspection of the failure surfaces.

The average failure loads and the average axial displacement at failure are summarized in Table 6.1 together with their standard deviation. The results in Table 6.1 show that the stitched single-lap joints are stronger and more flexible than the unstitched ones because they have greater strength and larger crosshead movement. For the long specimens with an unsupported length of 90 mm, through-thickness stitching leads to an average increase in joint strength by about 25% and in joint extension by about 28%. For the short specimens with an unsupported length of 70 mm, there is an average increase in joint strength by about 22% and in joint extension by about 31%.

Table 6.1 Effect of stitching on static failure strength of single lap joints fabricated by stitching preform and using RTM

Panel No.	Unsupported length (mm)	No. of specimens	Average failure load (kN)	Average cross-head movement at failure (mm)
1[a]	90 mm	7	11.22±0.41	2.36±0.12
2[a]	90 mm	7	11.45±1.02	2.73±0.29
3[a]	70 mm	8	12.37±1.12	2.69±0.31
4[b]	90 mm	8	14.11±0.76	3.26±0.23
5[b]	70 mm	8	15.06±0.77	3.52±0.32

[a] panel without transverse stitching. [b] panel with transverse stitching.

Fatigue test of a limited number of unstitched and stitched specimens are also conducted by applying a sinusoidal tension-tension load at a loading frequency of 3 Hz with a constant load ratio of 5. The specimens are tested to failure or 10^6 cycles. Figure 6.8 shows the results of the fatigue life of the stitched and unstitched specimens. Clearly, transverse stitching can improve the fatigue life by two orders of magnitudes for any given maximum tensile load. For a given cycle life, stitched specimens carry a significantly higher load than the unstitched specimens. In addition, for stitched joints, stable crack propagation along the interface between the two adherends is observed when the maximum load is only a fraction of the static strength of the unstitched specimens. The through-thickness stitches are found to bridge the cracked specimens.

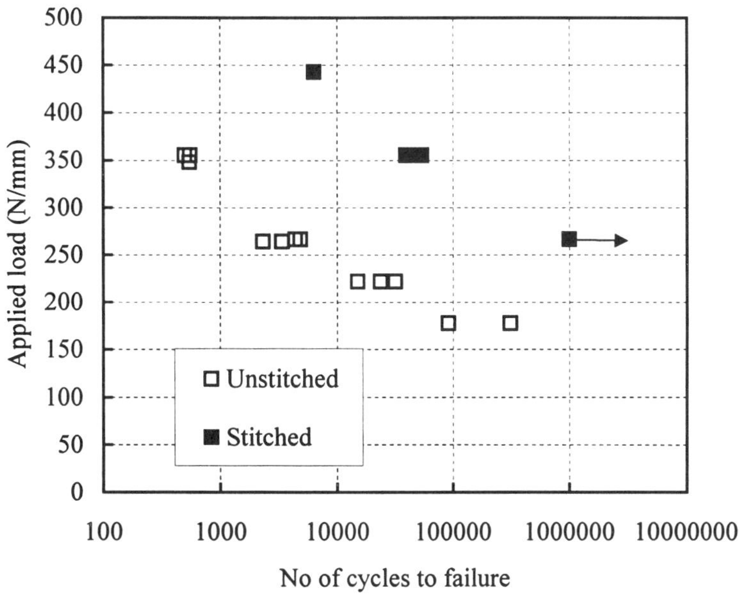

Figure 6.8 Effect of stitching on fatigue strength of single–lap joints loaded with a load ratio R=5 and at a frequency of 3 Hz

To better understand the effect of stitching and the RTM bonding, another two sets of static tests are performed for single lap joints. One is by bonding the adherends cut from the panel manufactured using the same RTM process using FM200-2K film adhesive (nominal adhesive thickness 0.12 mm)), and the other is to apply two types of clamp-up forces near overlap ends of unstitched single lap joints. Two clamp-up rigs are fabricated as shown in Figure 6.9. Figure 6.9(a) and (d) shows a clamp-up mechanism which has one row of loading contact at each end of the overlap, and Figure 6.9(b) and (e) depicts a clamp-up mechanism that has two rows of clamping contact of 3 mm apart at each end of the overlap. The clamp-up mechanism (see Figure 6.9(c)) allows application of a torque of 1 and 1.5 Nm by moving the loading arm towards the to the middle loading jack. Static tensile tests are performed at the same conditions as the unstitched and stitched joints.

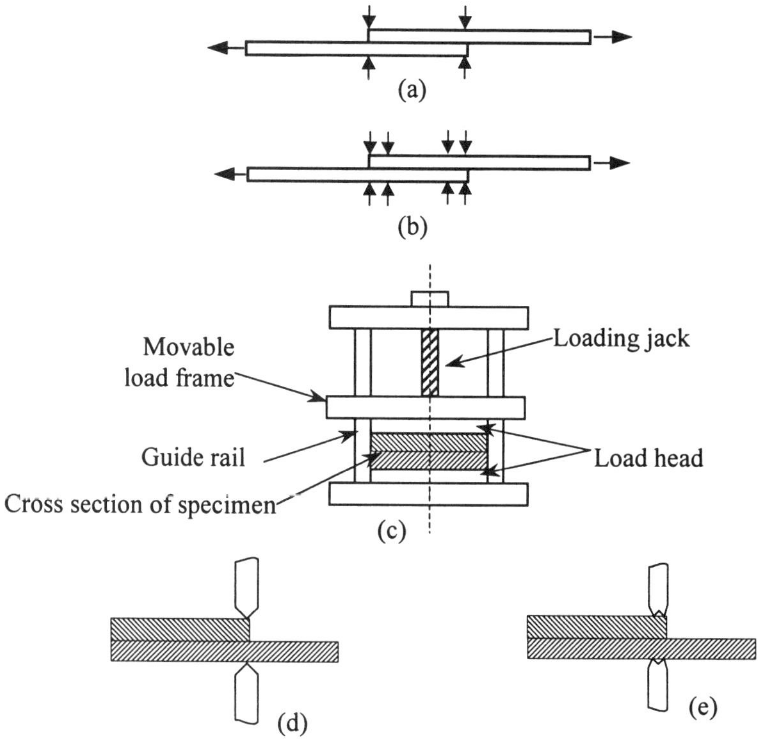

Figure 6.9 Joint set-up for applying transverse normal compressive forces

Table 6.2 compares the static failure loads of the unstitched, the stitched and the adhesive bonded joints as well as these with clamp-up mechanisms at both ends. A comparison of the failure load of the RTM unstitched (row 1 and 8) and adhesively bonded (row 2 and 9) specimens shows that both groups have almost the same failure loads. Inspection of failure loads in row 3 to 7 reveals that transverse stitching has the same effect of clamp-up forces. However, the actual induced clamp-up forces due to stitching may be lower than those due to the clamp-up mechanism with a torque of 1 and 1.5 Nm because the failure load of the stitched joints is slightly smaller than these clamped up with the mechanism shown in Figure 6.9. As yet there is no an available technique for measuring the effective clamp-up force caused by stitching in a cured composite joint.

Table 6.2 Measured failure loads (kN) of unstitched, stitched and adhesively bonded specimens with and without clamping forces

Specimen type	Clamp-up torque (N.m)	No. of clamp up rows	Average failure load
Unsupported length 90 mm			
RTM unstitched	/	/	11.33
Adhesively bonded	/	/	11.54
RTM stitched	/	/	14.11
RTM unstitched	1	1	15.27
RTM unstitched	1	2	14.89
RTM unstitched	1.5	1	15.55
RTM unstitched	1.5	2	13.95
Unsupported length 70 mm			
RTM unstitched	/	/	12.37
Adhesively bonded	/	/	12.17
RTM stitched	/	/	15.06

(b) Lap joints with thick adherends

Single-lap joint specimens were manufactured following the RTM process from the Hexcel Composites G926 EFP, 6k, 5 Harness satin weave carbon fabric and Ciba Araldite LY 556/ Hardener HY 917/ Accelerator DY 070 epoxy resin. All specimens had an overlap length of 30 mm, an unsupported length of 60 mm, and a width of 25.4 mm. The adherend had a lay-up of $[0/-45/45/90]_S$, and a nominal thickness of 2.864 mm. KEVLAR® 40 tex thread was used as the stitching thread. A zigzag stitch pattern was used at the overlap ends and a straight plain stitch pattern was used in the central region of the overlap with modified interlocking stitch as shown in Figure 6.10. Manufacturing and testing procedures were detailed in Tong and co-workers (1998).

Table 6.3 lists the average values of the final failure loads and the associated ultimate displacements as well as the percentage of the maximum deviations from the average values. The results of the failure loads of the specimens from Panel A are fairly consistent, and lie between

±7.2% of the mean. Panel B has failure loads varying between ±15.3 %. This is because the strength of the specimens cut from the centre of the panel were far below the mean ultimate strength of the panel. The center of the panel was of poor quality with surface porosity evident.

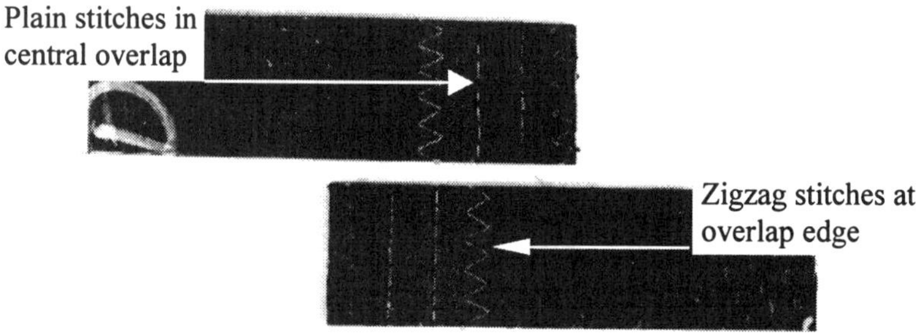

Figure 6.10 Photograph of single lap joint specimen with thick adherend

Table 6.3. Average failure loads and displacements for the specimens made from all panels

Panel No.	Failure load (kN)	Crosshead movement (mm)
A (unstitched)	9.74±7.2%	1.69±10.0%
B (unstitched)	9.52±15.3%	1.55±9.0%
C (stitched)	13.84±4.7%	2.21±9.5%
D (stitched)	13.91±3.7%	2.35±14.5%
E (stitched)	13.03±4.7%	2.13±8.5

Regardless of this less than perfect quality, Table 6.3 shows that there to be a significant increase in average strength of the stitched panel C, D and E comparing to that of the unstitched panel A and B. The average failure loads of panel C and D are over 42% higher than those of panel A and B. The highest increase in the strength is 46.1% for panel D comparing to panel B. This may be attributed to the high cure temperature applied to this panel. Panel E has an average strength increase of 33.8% and 36.9% comparing to that of panel A and B respectively. Panel E was of poor quality with higher porosity level compared to the other four panels. The overall average strength increase for all three stitched panels is 41.3% when compared to the unstitched panels.

The ultimate axial displacement given in Table 6.3 reveals that the stitched joints appeared to be more flexible than the unstitched in terms of having much higher ultimate axial displacement than the unstitched. For example, panel A has an average ultimate axial displacement of 1.69 mm, while panel C has an average ultimate axial displacement of 2.21 mm. The overall average increase in the ultimate axial displacement is 37.6%.

Although the materials and stitching patterns used are different for the two sets of specimens with thin and thick adherends, the testing results still give a clear indication that transverse stitching seems to be more effective in terms of increasing strength of a joint with thicker adherends.

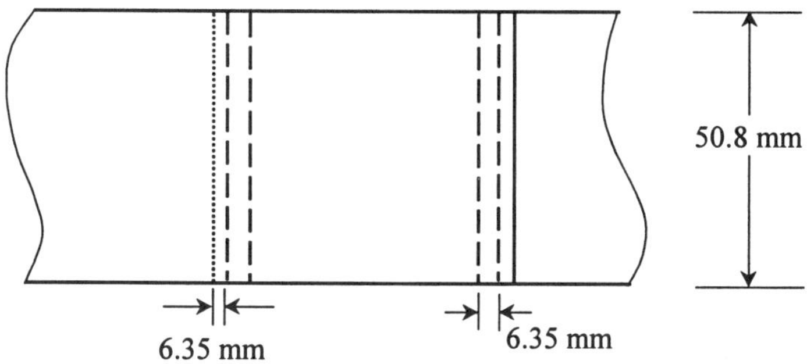

Figure 6.11 Top view of the four-row plain straight stitches used in the overlap of the single-lap joint by Sawyer (1985)

6.3.2 Joints fabricated by stitching prepreg and bonding

Sawyer (1985) presented an experimental study on the effect of stitching on strength of bonded composite single lap joints. The adherends were made of 16 and 24 plies of AS1/3501-6 graphite/epoxy with a quasi-isotropic lay-up. Specimens were prepared by stitching two panels sandwiched with a layer of FM-400 film adhesive using a shoe-making sewing machine and the Kevlar thread. Figure 6.11 schematically shows the four columns of plain straight stitches. The specimen width is 50.8 mm, the free length of the single lap joint is 127 mm, and the overlap length is taken to be 25.4, 50.8, 76.2 and 101.6 mm. Static and fatigue tests were performed.

Static test results show that the transverse stitching results in an up to 30% and 38% improvement in the joint failure load for the 16-ply and 24-

ply specimens, respectively, when compared to similar unstitched joints. Fatigue test results demonstrate that for a given peak load, the transverse stitching leads to at least one order of magnitude increase in the fatigue life of the stitched specimens as compared to the unstitched specimens. For a given fatigue cycle life, the stitched specimens carry a significantly higher load than the unstitched specimens.

6.4 STRENGTH PREDICTION

6.4.1 Solution method

The fracture mechanics approach is used in this section to predict the joint strength of the stitched and the unstitched single-lap joints with thin adherend described in section 6.3.1. For the single-lap joint subjected to axial tensile load as shown in Figure 6.6, let us consider the free body diagram of the overlap as depicted in Figure 6.12. Denote the adherend thickness as t, the overlap length as $2c$, and the unsupported length of the joint as l. The solution by Goland and Reissner (1944) gives the membrane forces and the bending moments as follows

$$N_i = P, \qquad M_i = k\frac{Pt}{2} \qquad (i=1,2) \tag{6.1}$$

where k is given in equation (2.13) in Chapter 2.

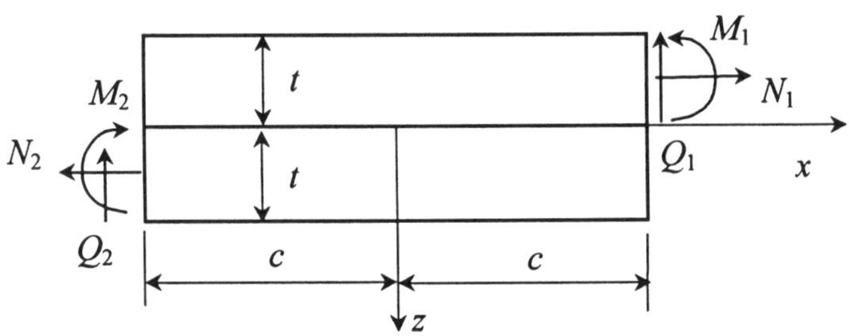

Figure 6.12 Stress resultants acting on the adherend at the overlap ends in a single-lap joint

The energy release rates of mode II and mode I at one overlap end can be calculated using the following equations:

$$G_{II} = \frac{A_1}{16}\left(\frac{N_i}{A_1} + \frac{t_1 M_i}{2D_1}\right)^2 \tag{6.2}$$

$$G_I = \frac{M_i^2}{4D_1} \tag{6.3}$$

where A_1 and D_1 are the longitudinal membrane stiffness and bending stiffness of each adherend. N_i and M_i (i=1,2) are the longitudinal membrane forces and the bending moments, as shown in Figure 6.12, at the overlap ends overlap acting on the top adherend at 1 and 2 respectively. It should be pointed out that the effect of transverse shear force is neglected in calculating the energy release rate.

Equations (6.2)-(6.3) indicate that the energy release rates at an overlap edge for a balanced single lap joint can be approximately determined in terms of the longitudinal force and the bending moment acting on the continuous adherend at an overlap end. The energy release rates can then be used to predict joint strength in conjunction with the following interactive linear failure criterion

$$\frac{G_{II}}{G_{IIC}} + \frac{G_I}{G_{IC}} = 1 \tag{6.4}$$

where G_{IC} and G_{IIC} are the critical energy release rate of mode I and mode II respectively that can be measured, for example, using a Double Cantilever Beam (DCB) test and an End Notched Flexure (ENF) specimen test.

An iterative numerical procedure is developed to predict the strength of the joints. In the analysis procedure, the applied load is increased gradually. For an applied load, the energy release rates of mode I and mode II are calculated using Equation (1-2) and then the failure criterion (equation (7)) is checked. When the failure criterion is satisfied, the corresponding load is regarded as the failure load of the joint.

Tong (1998) presents an equivalent modulus model to take into account the effect of stitching on the in-plane modulus. Figure 6.13 depicts the sketch of an ideal damage zone around a stitch, which is representative of the observation shown in Figure 6.7d.

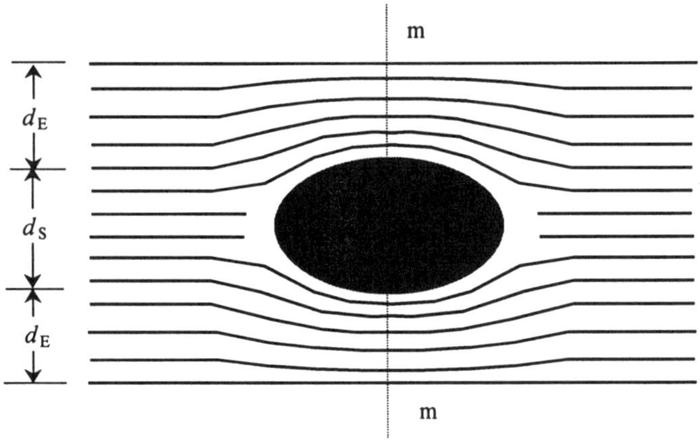

Figure 6.13 A sketch of the damage zone around a single stitch

In this model, it is assumed that the Young's modulus in the fibre direction in the damage zone is degraded to the Young's modulus in the transverse direction (perpendicular to the fibre direction), while the Young's modulus in the undamaged part between two stitching holes remains unchanged. Along the dotted line *mm* as shown in Figure 6.13, consider a parallel spring model consisting of a stiff spring with a stiffness of E_{11} , which represents the undamaged part, and a soft spring with a stiffness of E_{22}, which represents the damaged part. The volume fraction of the damaged part is defined as the ratio of the measured width of the elliptical stitch shape to the stitch length l_{SD} in the direction of the specimen width. Hence the degraded Young's modulus in the fibre direction is given by

$$E_{11}^d = E_{11}(1 - v_d) + E_{22}v_d \tag{6.5}$$

where $v_d = (d_S + 2d_E)/l_{SD}$.

Tong (1998c) proposed an effective stitch density model based on the visual observation of the stitch loops near overlap edge. Figure 6.14(a) shows an ideal rectangular stitch loop near an overlap edge. However, due to existence of pretension in stitch and low stiffness and strength of preform, most stitches cut into the dry preform and ideal rectangular stitch loops could not be maintained. The distorted stitch loop is shown schematically in Figure 6.14(b). However, as shown in Figure 6.14(b), the distorted stitch loop right at the overlap edge can not achieve the full clamping force as an ideal rectangular stitch does. Comparing to the full clamping force F_C provided by an ideal stitch loop, the clamping force F_{CD} of a distorted stitch loop is assumed to be only a fraction of the full clamping force. As an approximation, the pretension force in an ideal stitch loop is assumed to be the same as that in a distorted stitch loop. Hence the clamping force is equal to the pretension force in an ideal stitch loop, while the claming force in a distorted stitch loop can be approximated as the pretension force multiplied by a factor of $\sin\theta$.

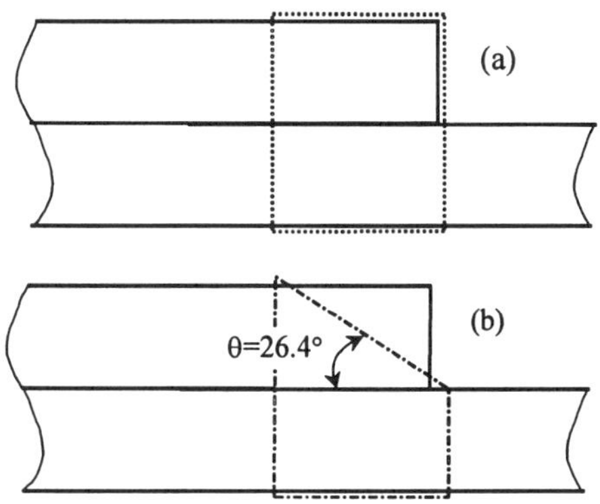

Figure 6.14 Stitch loops close to an overlap end (a) an ideal stitch loop, (b) an observed stitch loop

6.4.2 Results and discussion

The single lap joints presented in Section 6.3.1 have an overlap length of 30 mm and a width of 25.4 mm. The free length L is 90 mm and 70 mm respectively. The specimens have a lay-up of $(0/\pm45/90)_s$ for each adherend, and each ply had a nominal thickness of 0.205 mm. Tables 6.4 and 6.5 list the mechanical properties of a ply and the critical energy release rates respectively (Jain et al,1995; Tong, 1998c).

Table 6.4 Ply mechanical properties of unstitched RTM laminate

Longitudinal modulus E_1 (GPa)	99.8
Transverse modulus E_2 (GPa)	36
Shear modulus G_{12} (GPa)	5.8
Major Poison's ratio ν_{12}	0.33

Table 6.5 Critical energy release rates G_{IC} and G_{IIC} of unstitched and stitched composite laminates

Stitch density (st/cm^2)	G_{IC} (kJ/m^2)	G_{IIC} (kJ/m^2)
0	0.34	1.2
4	0.62	2.2

Table 6.6 compares the average measured failure loads per unit width with the predicted failure loads. The failure loads of the joints with L=70 mm are predicted to be the same as those of the joints with L=90 mm. The average measured failure loads of the joints with L=90 and 70 mm are 7.7% lower and 0.8% larger than the predicted failure loads respectively. Evidently there is a good correlation between the average measured and the predicted failure loads.

Table 6.6 Failure loads (N/mm) of the unstitched specimens

Group No.	Unsupported length	Measured	Predicted
1	90	441.6	483.5
2	90	451.0	483.5
3	70	487.2	483.5

For the stitched joints, the effective longitudinal modulus is found be 80.66 GPa and the effective stitch density is determined as to be 2.625 stitches/cm^2. Using a linear interpolation, the critical energy release rates of mode I and II corresponding to a stitch density of 2.625 stitches/cm^2 are evaluated as 0.524 and 1.856 (kJ/m^2), respectively. Following the solution procedure, the failure loads of the stitched single lap joints are determined and are given in Table 6.7. The failure loads for the stitched joints with a free length of 90 and 70 mm are predicted to be 585 N/mm. The measured failure loads of the joints having a free length of 90 and 70 mm are respectively 5% lower and 1.4% larger than the predicted failure loads. This result shows a good correlation between the measured and predicted failure loads.

Table 6.7 Failure loads (N/mm) of the stitched specimens

Group No.	Unsupported length	Measured	Predicted
4	90	555.7	585.0
5	70	593.1	585.0

6.5 NUMERICAL MODELING AND ANGLED STITCHING

The experimental results in section 6.3 have shown that through-thickness stitching can significantly improve the strength of the single-lap joints manufactured using the RTM procedure. Although joint strength can be predicted using the solution based on fracture mechanics, little is known about the nature of the stresses in a RTM lap joints with and without the through-thickness stitching. In this section, the numerical examples given by Tong and Steven (1996) are presented to demonstrate the finite element modeling schemes, and some basic features of the stress distributions in the unstitched and stitched RTM joints. The effect of the stitching angle on the stress distribution in the adherends and the axial forces in the stitching thread is also demonstrated.

6.5.1 Finite element modeling

Consider the single-lap joint with some of its dimensions as shown in Figure 6.15. The joint is assumed to be manufactured following the RTM process. The two adherends are formed by stacking a number of preform plies in the designated directions, and then they are overlaid for a certain length and finally transversely stitched together using a modified lock stitch at the overlap ends prior to the injection of resin. Unlike an adhesively bonded joint, there is no adhesive layer between the two adherends for RTM joints although interlaminar cohesion is the load transfer mechanism.

Modeling of stitched joints depends on many parameters, such as stitching pattern, material properties of stitch thread, stitch diameter, in addition to the most commonly used parameters for bonded joints. Among these stitching pattern is very important, it consists of stitching density, pitch length, stitch length as well as stitch path. At structural level, a stitch thread can be modeled as a beam, thus a beam and brick element combined modeling scheme becomes a straight forward modeling strategy. In this case, stitch thread path is an extremely important modeling aspect and also very difficult. The beam can be modeled using normal 3D beam elements or can be approximately modeled as 3D beam capable of sustaining tension or compression only (with no bending stiffness). As the bending stiffness of a stitch thread can be much smaller than its extensional stiffness, it is reasonable to neglect the effect of bending. In this section, a simplified plane strain analysis will be presented to demonstrate the effect of through-thickness stitching.

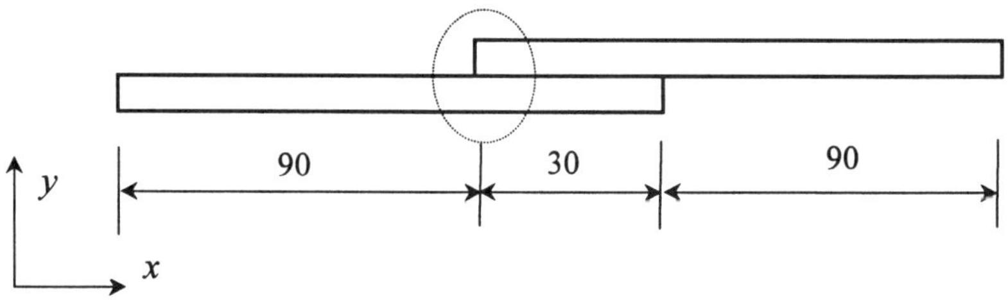

Figure 6.15 Configuration for a single lap joint subjected to a tensile load

It is assumed that the left end of the joint in Figure 6.15 is fully clamped while the right end is allowed to move in the load-path direction only, e.g., 0.5 mm in the *x* direction and 0.007 mm in the y direction. As the joint is 25.4 mm wide, it is reasonable to assume that the FEA model of the joint can use plane strain elements. A geometrically nonlinear 2-D plane strain analysis is carried out using the Strand6 finite element analysis system to investigate the effects of some stitching parameters on the nature of the stresses in the single-lap joints.

Modeling of laminated adherends: The lay-ups for the two adherends are assumed to be (0/90/90/0)s. Material properties used in the analysis are: E_L=127.5 GPa, E_T=9.0 GPa, E_Z=4.8 GPa, $\nu_{LT}=\nu_{LZ}$=0.28, ν_{TZ}=0.41, $G_{LT}=G_{LZ}$=4.8 GPa, and G_{TZ}=2.25GPa. The thickness of each ply is 0.1875 mm. 2-D four-node quadrilateral plane strain elements are used to model the adherends using a ply-by-ply meshing scheme. Figure 6.16 depicts the mesh used near an overlap end. A fine mesh is used at both ends of the overlap with elements of dimensions, 0.1875 mm by 0.2222 mm.

Geometric nonlinearity is included in the analyses to account for the large rotation and deflection occurred in the overlap region.

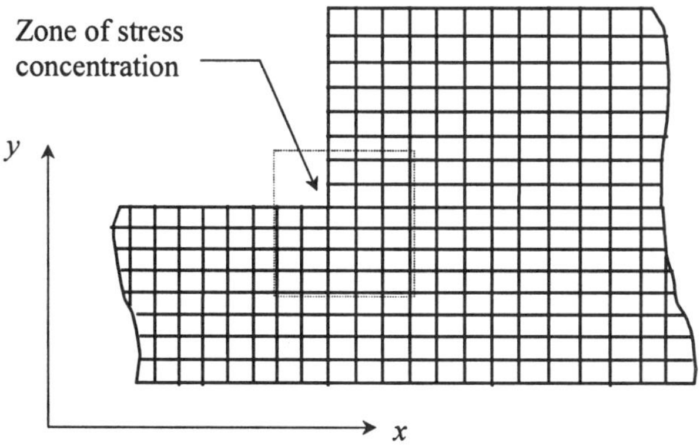

Figure 6.16 Meshing scheme at one end of the lap joint using 4-node quadrilateral plane strain elements

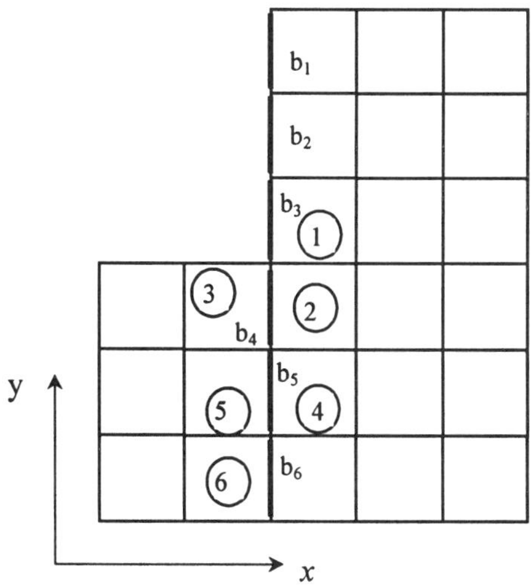

Figure 6.17 Four-node plane strain and beam elements at an overlap end of the joint reinforced with vertical stitching (elements taken from the marked area in Figure 6.16)

Modeling of stitches: The stitch thread is assumed to carry axial tensile or compressive load only, and thus can be modeled by tension-compression only beam elements or bar elements of circular cross section. The diameter of the stitch thread is taken to be 0.2 mm. The baseline value of the tensile and compressive modulus of the stitch thread is taken to be 120 GPa. The modulus of stitch thread is then varied from 1 to 160 GPa to study its effect on the nature of the stresses. The extensional stiffness of the beam element is computed by doubling the extensional stiffness of one stitch thread as each transverse stitch consists of two stitch threads in the modified lock stitch. Figure 6.17 depicts some of the stitch elements and the plane strain 4-node quad elements in the stress concentration zone.

All calculations are carried out using the Strand6 nonlinear solver with 5 load steps and 20 iterations in each load step. All results are obtained by fully clamping the left end and moving the right end outward by 0.5 mm in the *x* direction. The corresponding vertical movement in the y-direction is imposed using the constraint function with values obtained from the actual load-path direction of the joints.

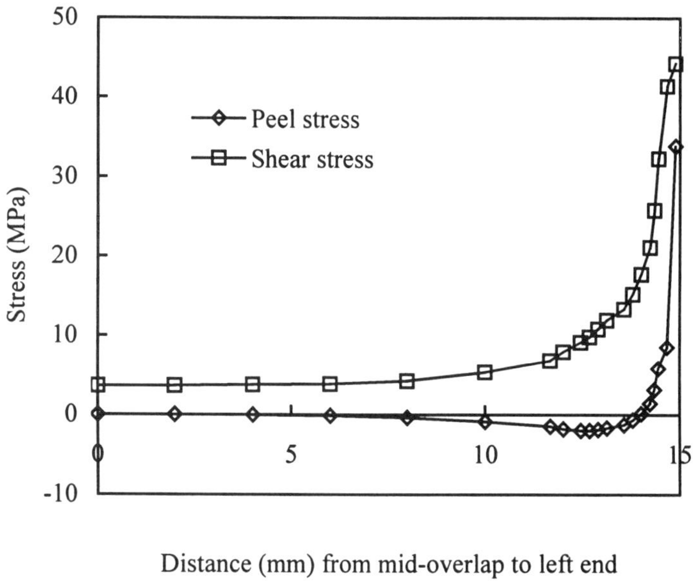

Figure 6.18 Distribution of the peel or normal stress σ_{yy} and shear stress σ_{xy} in surface ply of the unloaded adherend measured from the mid-point of the overlap to the left end of the joint in Figure 6.16

6.5.2 Stresses in the unstitched joint

To study the effects of the transverse stitching parameters, an unstitched lap joint is analyzed first to establish a benchmarking reference. The unstitched lap joint will be referred to as Case 1 in the following discussion. Inspection of the stress contours reveals that the peel and shear stress concentrations take place at the ends of the overlap, and the peel and shear stresses in the central region of the overlap are relatively small as shown in Figure 6.18. This is similar to the stress distributions found by many researchers for an adhesive bonded single lap joint. This clearly indicates that the transverse stitching, which provides a clamping force or stress mechanism, should be applied at the ends of the overlap to lessen the stress concentrations, especially the peel stress. The three critical elements are element 1, 2, and 3, as shown in Figure 6.17. The stresses in these three elements will be used as the baseline values for demonstrating the stitching effects.

6.5.3 Stresses in the stitched joints

(a) Effect of the pitch length

Pitch length is defined as the distance between any two adjacent stitching holes, and it is an important stitching parameter in the stitching process. Transverse stitching introduces a reinforcing loop, as shown in Figure 6.14, and in the loop there is only one single stitch thread on the top and bottom surfaces and there are two stitch threads running through the thickness of laminates.

To understand the effect of pitch length on the critical stresses, analyzed are the joints with three different stitching pitch length, i.e. 1.33 mm, 2.67 mm and 4.0 mm, respectively. The modulus of the stitch is taken to be 120 GPa. One of the two through-the-thickness stitches is taken to be located right at the end of the terminating adherend.

Table 6.8 gives the longitudinal stresses σ_x, peel stresses σ_y and the shear stresses τ_{xy} in element 1, 2 and 3, as shown in Figure 6.17, for the unstitched and the stitched joints with three pitch lengths. Comparison of the stresses in both unstitched and stitched joints shows that transverse stitching significantly decreases the peel stresses in all three elements. This demonstrates the important effect of transverse stitching in terms of reducing the critical peel stress that is regarded as a major cause of delamination. In element 1 (see Figure 6.17), the transverse stitching increases the shear stresses and decreases the longitudinal stresses, while in element 2 and 3 (also see Figure 6.17) the transverse stitching increases the longitudinal stresses and reduces the shear stresses. However, when comparing the stresses in the three critical elements for the three pitch lengths, i.e., 1.33 mm, 2.67 mm and 4mm, it is found that they remain almost unchanged. This result clearly shows that the pitch length does not affect the stresses in the three critical elements once it is larger than 1.33 mm. This is similar to the conclusion obtained using an analytical approach by Tong and Jain (1995).

Table 6.8 Effect of the pitch length on stresses (MPa) in the three critical elements

Pitch length	σ_x	σ_y	τ_{xz}
	Element 1		
Case 1*	159.0	33.7	43.9
1.33 mm	154.0	6.4	51.2
2.67 mm	154.9	6.2	51.6
4 mm	155.2	6.3	51.7
	Element 2		
Case 1*	448.2	24.0	16.7
1.33 mm	449.1	7.4	11.7
2.67 mm	450.0	7.2	11.9
4 mm	450.1	7.3	11.9
	Element 3		
Case 1*	607.4	19.9	38.3
1.33 mm	610.9	10.2	34.2
2.67 mm	612.2	10.2	34.5
4 mm	612.7	10.3	34.7

* unstitched lap joint

Table 6.9 presents the axial forces in the six beam or stitch elements (see Figure 6.17) which indicate the tensile forces in the stitch threads close to stress concentration zone. The tensile force in the stitch attains maximum value in element 4 which indicates that the axial tensile stresses are 315, 312 and 315 MPa respectively for the joints with three pitch lengths. The critical peel stress σ_y in element 1 is reduced from 33.7 MPa of unstitched case to about 6.4 MPa of the stitched case by stressing the stitch element 4 with about 315 MPa. This demonstrates that the stitch plays an important role in the load transferring between the two adherends. The tensile stresses carried by the stitch elements are very close to each other for the stitched joints with three pitch lengths, and this indicates that varying pitch length of the stitching loop does not have noticeable effect on the stresses in the transverse stitch. Experimental results (Tong, Jain, et al, 1998) showed that the stitches close to the end of the overlap broke inside the continuous adherend and then were pull out. This observation may be interpreted in terms of the numerical results given in Table 6.9. From this table, it can be seen that the tensile force in the stitch elements reaches the maximum value stitch element 4 (see Figure 6.17). Thus this result clearly supports the argument that the stitch breakage takes place in the continuous adherend rather than in the terminated adherend.

Table 6.9 Effect of pitch-length on the axial forces (N) in the six stitch elements

Pitch length (mm)	b1	b2	b3	b4	b5	b6
1.33	2.0	3.9	9.4	9.9	6.8	5.2
2.67	1.8	3.8	9.3	9.8	6.7	5.1
4	1.9	3.9	9.5	9.9	6.8	5.3

(b) Effect of the Young's modulus of stitch thread

To study the effect of stitch modulus, the lap joint is modeled with one vertical stitch thread at the edge of the terminating adherend. The modulus of stitch takes values of 1, 10, 40, 80, 120 and 160 GPa respectively. As above, stress concentration occurs at the end of the overlap in the three critical elements, i.e., element 1, 2 and 3 of Figure 6.17. The stresses in the three elements are given in Table 6.10. While increasing the stitch modulus has negligible effect on the stress σ_x, and it does significantly change the peel stress σ_y and the shear stress τ_{xy} . Increasing stitch modulus from 1 to 120 GPa reduces the peel stress σ_y from 33.7 to 5.2 MPa in element 1, from 24.0 to 6.1 MPa in element 2 and from 19.9 to 9.4 MPa in element 3. However, increasing stitch modulus from 80 to 160 GPa (which is typical range of stitch modulus, such as Kevlar, carbon, etc) tends to cause less change in the peel stress.

Figure 6.19 gives the axial forces in the six stitch elements that represent the tensile forces in the stitch close to the stress concentration zone. The maximum stitch tension occurs in stitch element 3 for the stitched joints with low stitch modulus and in stitch element 4 for the joints with high stitch modulus. The axial tensile forces in the six stitch elements increase as the stitch modulus is enlarged, and thus the load transferred through the stitch elements increases. As the load passing through the stitch elements becomes larger, the peel stresses transferred through the three elements are lowered. When the stitch modulus is larger than 80 GPa, varying stitch modulus tends to cause less change in the tensile forces in the stitch elements.

Table 6.10 Effect of stitch modulus on stresses (MPa) in the three critical elements

Modulus of stitch thread (GPa)	σ_x	σ_y	τ_{xz}
	Element No 1		
Case 1*	159.0	33.7	43.9
1	158.6	31.2	44.5
10	157.1	20.4	47.3
40	155.4	11.2	49.7
80	154.5	7.7	50.6
120	154.0	6.1	51.0
160	153.7	5.2	51.2
	Element No 2		
Case 1*	448.2	24.0	16.7
1	448.1	23.3	16.1
10	447.8	18.6	13.3
40	448.1	11.9	11.8
80	448.5	8.7	11.7
120	448.6	7.1	11.7
160	448.8	6.1	11.8
	Element No 3		
Case 1*	607.4	19.9	38.3
1	607.6	19.6	38.4
10	608.6	17.4	38.2
40	609.7	13.3	36.3
80	610.0	11.1	34.9
120	610.1	10.1	34.0
160	610.1	9.4	33.4

* unstitched lap joint

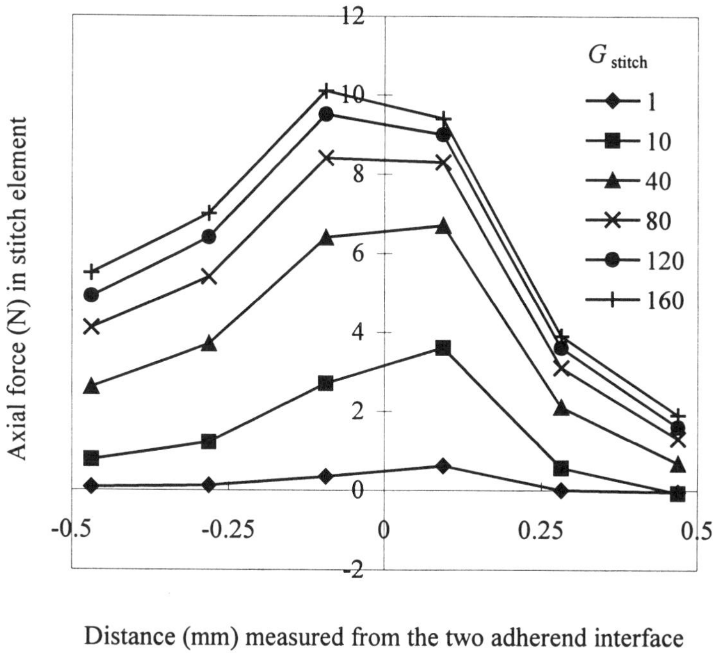

Figure 6.19 Effect of stitch modulus G_{stitch} on the axial force distribution in the stitch thread segment near the interface between the two adherends

(c) Effect of the off-edge distance

In studying the effect of stitch modulus, the stitch threads run through the thickness right at the edge of the terminating adherend. In practice, although care can be taken to place stitching as close to the edge as possible, it is rather difficult if not impossible to stitch right at the edge without reshaping the reform of the terminated adherend. Therefore it is of particular important to study the effect of off-edge distance on the critical stresses in the lap joints. The off-edge distance is taken to be 0.22, 0.44 and 0.67 mm respectively. The stitch modulus is 120 GPa. Similar to the unstitched joint, stress concentration occurs in the three elements as shown in Figure 6.17. The stresses in the three elements are listed in Table 6.11.

It is seen that there is a negligible effect on the longitudinal stresses σ_x as the stitch thread is moving away from the edge of the terminated adherend. However, the off-edge distance has a dramatic effect on the peel stress σ_y

and a noticeable effect on the shear stress τ_{xy}. The peel stress in element 1 is reduced when the stitch is close to the edge. This stitching effect disappears when the stitch thread is about 0.67 mm off the edge of the terminated adherend. This observation can also be demonstrated by the axial forces in the six stitch elements in Figure 6.20. In this figure it is observed that the axial forces in the stitch elements 2 to 6 become small and even negative when the off-edge distance increases from zero to 0.67 mm. When the off-edge distance is over 0.44 mm, the axial forces in element 4, 5 and 6 (see Figure 6.17) become negative. This indicates that the transverse stitching is not necessary as it runs through a zone that has compressive stresses.

Table 6.11 Effect of the off-edge distance on stresses (MPa) in the three critical elements

Off-edge distance (mm)	σ_x	σ_y	τ_{xz}
	Element No 1		
Case 1*	159.0	33.7	43.9
0.0	154.0	6.1	51.0
0.22	161.2	27.3	45.0
0.44	159.9	31.8	44.4
0.67	159.1	33.7	44.2
	Element No 2		
Case 1*	448.2	24.0	16.7
0.0	448.6	7.1	11.7
0.22	448.3	21.2	17.0
0.44	448.1	23.5	17.5
0.67	448.0	24.4	17.2
	Element No 3		
Case 1*	607.4	19.9	38.3
0.0	610.1	10.1	34.0
0.22	607.9	18.4	37.3
0.44	607.5	19.7	38.3
0.67	607.2	20.0	38.4

* unstitched lap joint

Figures 6.21 and 6.22 compare the peel and shear stress distributions along the surface ply of the upper adherend of the joint in Figure 6.15. It is noted that both peel and shear stresses have the same distribution patterns as these found in an adhesive bonded single lap joint. However, stitching significantly reduce the peak peel stress.

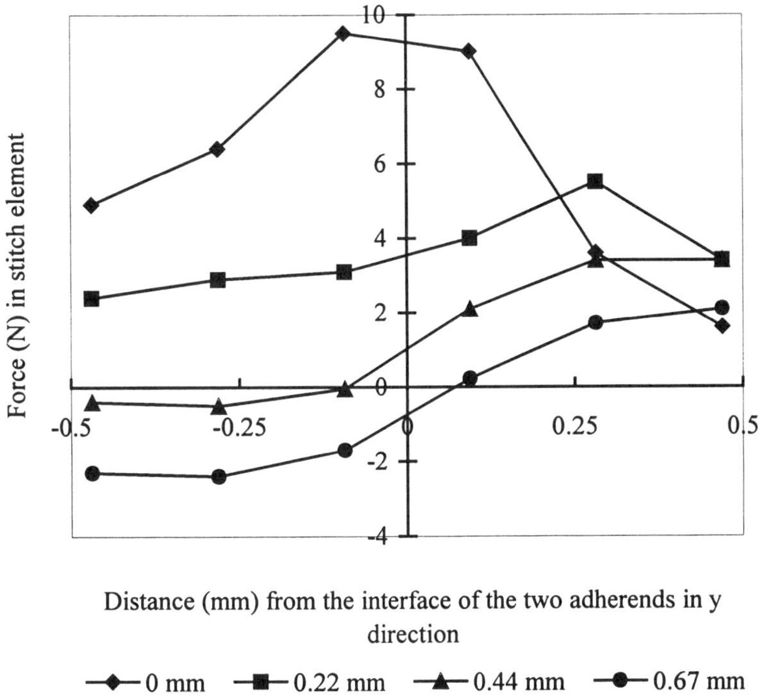

Figure 6.20 Effect of the off-edge distance on the axial force distribution along the stitch thread segment near the interface between the two adherends

The above numerical results demonstrate the effects of the stitch pitch length, stitch modulus and the off-edge distance on the stresses in the stitched joints. They clearly show that all these three parameters can reduce significantly the critical peel stresses in all three elements. However they may simultaneously increase or decrease the related shear and longitudinal stresses. While the peel stress is generally regarded as an important factor in causing delamination in composites, the shear stress can also be a contributing factor in the form of interacting with the peel stress as described in Chapter 3.

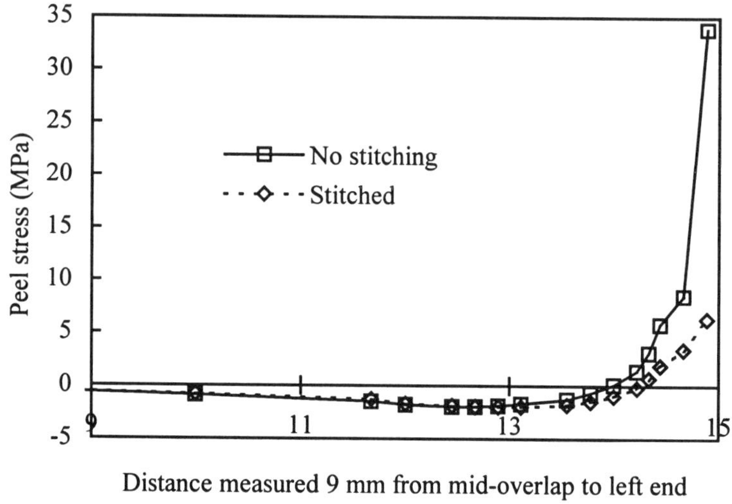

Figure 6.21 Effect of stitching on shear stress distribution in the surface ply of the upper adherend of the joint in Figure 6.15

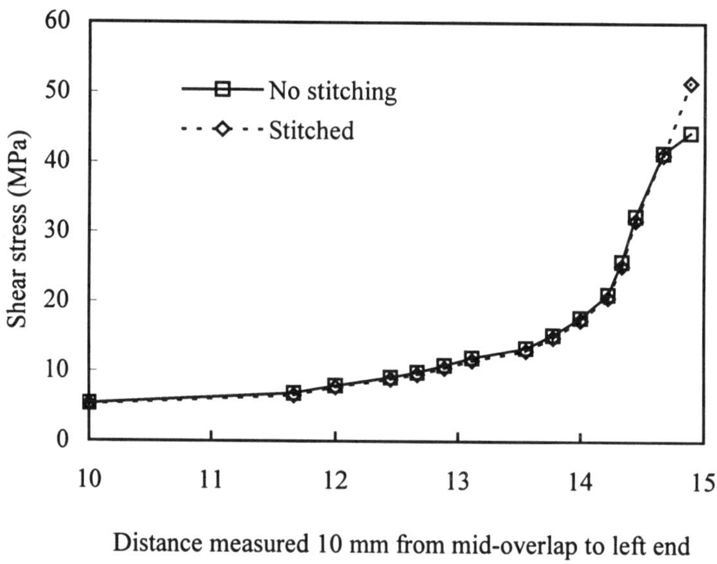

Figure 6.22 Effect of stitching on shear stress distribution in the surface ply of the upper adherend of the joint in Figure 6.15

It is desirable to reduce the shear and longitudinal stresses while decreasing the critical peel stress in the stress concentration zone. From the above numerical analyses, it is clear that it is difficult to achieve this by following the traditional vertical stitching process. In the following section, we will numerically demonstrate that angled stitching can reduce both the peel and shear stresses.

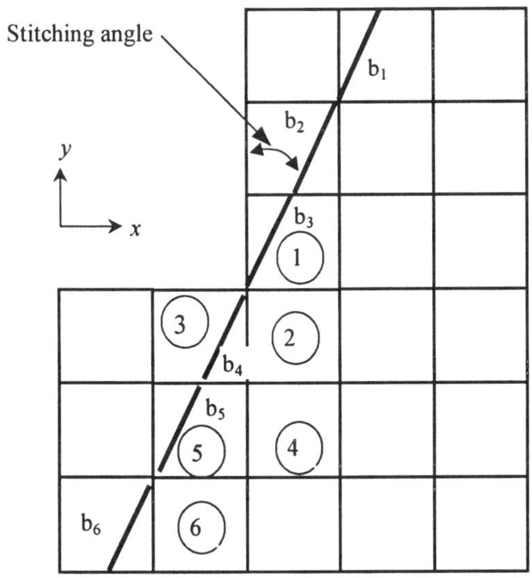

Figure 6.23 Four-node plane strain and beam elements at an overlap end of the joint reinforced with angled stitching (elements taken from the marked area in Figure 6.16)

6.5.4 Angled stitching

The direction of the principal stress in adhesive is always at an angle to the through-thickness direction (Adams and Wake, 1984). To effective reduce the peak stresses in adhesive, it is naturally desirable to place stitches in or close to the direction of the principal stress. The results of unstitched RTM lap joints show the similar results. In this section, the effect of the angled fibre stitches on peak stresses will be demonstrated for the single lap joint shown in Figure 6.15. To verify the new concept, the following stitching angles 8.4°, 16.5°, 30.7° and 49.8°, which facilitate the mesh in

Figure 6.6, are selected to perform a parametric study. The stitching angle is defined as the angle between the y-axis and the stitch as shown in Figure 6.23, which depicts the detailed mesh used and the critical elements near an overlap end where stress concentration occurs.

All solution conditions are the same as those explained in the previous vertical stitching section except that constraint conditions are imposed at the ends of the beam element in terms of the adjacent two nodal displacements of the 4-node quadrilateral element.

Table 6.12 Effect of stitch angle on stresses (MPa) in the three critical elements

Stitch angle	σ_x	σ_y	τ_{xz}
	Element No 1		
Case 1*	159.0	33.7	43.9
0°	154.0	6.1	51.0
8.4°	156.4	0.15	47.3
16.5°	157.9	-4.5	42.6
30.7°	153.0	-7.5	31.4
49.8°	128.2	6.1	22.1
	Element No 2		
Case 1*	448.2	24.0	16.7
0°	448.6	7.1	11.7
8.4°	444.8	3.5	8.8
16.5°	437.7	0.87	5.5
30.7°	420.8	0.18	2.0
49.8°	405.9	7.9	5.8
	Element No 3		
Case 1*	607.4	19.9	38.3
0°	610.1	10.1	34.0
8.4°	608.4	7.2	30.7
16.5°	604.0	4.4	26.6
30.7°	585.0	1.5	17.1
49.8°	539.6	5.4	10.0

* unstitched lap joint

As shown from the results for the cases of unstitched joint and vertically stitched joints, one of the three elements (element 1, 2 and 3 in Figure

6.17) always has the largest peel stress. However, for the case of non-vertically stitched lap joints, the maximum peel stress no longer occurs in these three elements and it takes place in element 5 (see Figure 6.23). For comparison, the stresses in the three elements for the unstitched, vertically-stitched and non-vertically stitched joints are listed in Table 6.12. It is shown that, while vertical stitching reduces the peel stress, angled stitching can also decrease the peel stresses and even create negative peel stresses. It is evident that when the stitch angle is 16.5° and 30.7°, all stresses in the three elements of the non-vertically stitched joints are reduced when comparing to those of the unstitched and vertically-stitched lap joints. However, reduction in the longitudinal stresses is small comparing to that in the peel and shear stresses. Results in Table 6.12 clearly indicates that the stitch angle is another stitching parameter that can reduce all the stresses in the stress concentration zone, element 1, 2 and 3.

Table 6.13 Stresses in the element that has the maximum peel stress (MPa) for different stitch angles

Stitch angle	Element	σ_x	σ_y	τ_{xz}
Case 1*	1	159.0	33.7	43.9
0°	5	26.7	10.4	5.8
8.4°	5	26.7	9.5	3.9
16.5°	5	26.3	9.8	1.2
30.7°	5	26.4	13.8	-2.6
49.8°	5	24.0	12.4	-6.3

* unstitched lap joint

It is found that the peak peel stress occurs in element 5 of Figure 6.23 for lap joints with angled stitches. Table 6.13 lists the three stresses in element 5. Note that the lowest peel stress is 9.5 MPa when the stitch angle is 8.5°, and the second lowest one is 9.8 MPa when the stitch angle is 16.5°. If the high peel stress is the primary contributing factor to the interlaminar delamination, based on the results in Table 6.12 and 6.13 and the maximum stress criterion, then a stitching angle of approximately 16.5° could be regarded as the optimum one in terms of maximizing the static failure load. If an interactive failure criterion as given in section 3.2 is used, due to the contribution of the longitudinal stress, optimum stitching angle becomes dependent on both interlaminar and longitudinal strengths as well as the stresses in element 1, 2, 3 and 5.

Figure 6.24 presents the axial forces in the middle of the six inclined beam or stitch elements as shown in Figure 6.23. The axial forces in all stitch elements are in tension and increase with the stitch angle. For the joints with an stitch angle of 8.4°, 16.5°, 30.7° and 49.8°, the axial tensile forces in stitch element 4 are respectively 29%, 58%, 111% and 155% larger than that for the vertically stitched joint as shown in Figure 6.19. These results clear show that: when the stitch angle increases, the stitching threads transfer more loads, which in turn results in a smaller amount of load being transferred through the weak resin interface between the two adherends. Thus angled stitching is more effective than vertical stitching in terms of stress diffusion as the stronger stitching threads carry more load than the weak resin matrix.

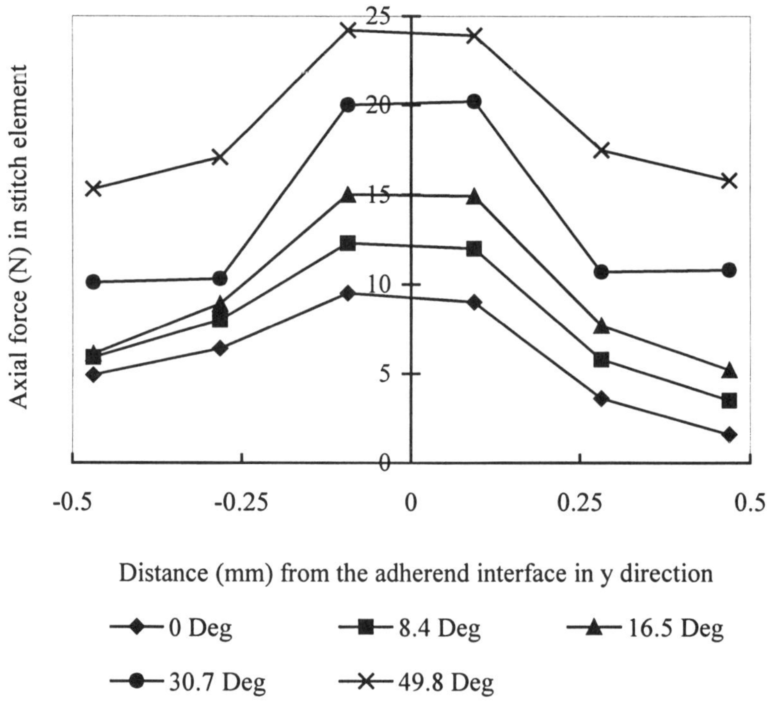

Figure 6.24 Effect of the stitch angle on the axial force distribution along the stitch thread segment near the interface between the two adherends

Numerical results demonstrate that angled stitching can reduce all three stresses in the stress concentration zone of the single lap joints. Angled stitching provides a new means of reducing significantly the peak peel and shear stresses in the critical peel stress concentration zone, which is generally regarded as a major contributing factor to delamination. The new stitching concept should be incorporated into a fully computer-controlled industrial sewing machine that can provide transverse stitching at various angles. Angled stitching has the potential to be used in manufacturing large complex structural components with thickness discontinuities, and it may be used to overcome the limitations of 3D orthogonal and angle interlocked woven composite materials.

7

Selected applications

7.1 INTRODUCTION

Adhesives have found many uses in a great variety of industries, for example, aerospace, automobile, marine, civil, sports, dental etc. It is easy for everyone to identify examples of adhesive bonding in the world around us. Because of this widespread usage it is beyond the scope of this text to identify and document all of the applications of adhesives even in engineering assembly and fabrication. This chapter aims to briefly review some of the major applications of adhesive bonding in some engineering sectors, such as aerospace, automobile, marine and civil industries. For further detailed information, keen readers are referred to the books by Adams, Comyn and Wake (1997), Mays and Hutchinson (1992) and the articles by Kinloch (1997), Hart-Smith and Strindberg (1997).

Applications of adhesive bonding in the various industry sectors can be grouped into two general categories: non-structural and structural (load-carrying) applications. For example, non-structural applications include gap filling and bonding on trim for decorative purposes. Structural applications include secondary bonded stiffened structures with a variety of cross-sections of stringers, sandwich structures formed by bonding thin sheets of materials onto various core materials. Bonded repair represents another major application whereby structural integrity (strength and stiffness) is restored thereby achieving cost-effective structural maintenance.

7.2 AEROSPACE

The use of adhesives in aerospace has a distinguished technical record and goes back a very long and interesting way. Some of the earliest flying machines used adhesives extracted from natural materials, such as beeswax and casein. These naturally occurring polymeric adhesives worked well except when exposed to hostile environments, such as heat, moisture or high stress. Synthetic polymeric adhesives were introduced to overcome the poor aging problem of adhesives based upon natural materials. One successful application of structural adhesives was made during World War II for bonding the wooden structure of the very successful 'Mosquito' aircraft.

The early synthetic adhesives for bonding wooden and metallic structures were rather brittle in nature and thus prone to cracking and fracturing. In early 1940, Ciba Geigy introduced a tougher adhesive Redux 775 for bonding metal honeycomb to metal skins. This toughened adhesive was extensively used by the engineers at the de Havilland Company in the UK for structural bonding in the Comet jet airliner. This adhesive has a good and well-proven track record and is still in use today on some Airbus aircraft.

In 1948 the SAAB J29 aircraft only had 10 adehsively bonded metal parts, and by 1967 there were 330 bonded parts on the SAAB AJ37. In the late 1970s, the Primary Adhesively Bonded Structure Technology (PABST) research programme was performed at Douglas Aircraft under contract to the US Air Force Wright Laboratories. The PABST wide-body fuselage was 12.8 meters long and 5.5 meters diameter and it was designed by analyzing and sizing adhesive bonded joints and then manufactured using subsequently developed tooling technology. Tooling technology has been further developed to produce bonding of the stiffened wings skins for the SAAB 340 and 2000 aircraft. These aircraft probably represent the most advanced and most challenging application for large-scale metal bonding (Hart-Smith and Strindberg, 1997). On the SAAB 340 aircraft, bonded parts include the entire skin of the constant section, parts of the curved nose and tail of the fuselage, the stiffened skins of the main wing box and the sandwich skins of the horizontal and vertical tails.

It can be said that the aerospace industry pioneered the application of adhesives in structural bonding, as well as many other technological innovations. Today adhesives are used to bond and repair critical components in commercial and military aircraft, helicopters and spacecraft. In the aerospace industry, safety, reliability and cost command paramount attention. These features are also the driving force for most technological innovations. The motivation to use adhesive bonding stems from the desire to achieve high stiffness and strength-to-weight ratios whilst at the same time prolonging aircraft life. Compared to the mechanical fastening, the advantages and disadvantages have been discussed in Table 1.1 in Chapter 1. The airframes of modern aircraft use a combination of materials, such as aluminum alloys, fiber reinforced composite materials, titanium and stainless steels, to achieve high stiffness- and strength-to-weight ratios. Examples include the Boeing 747, 767 and 777, Airbus 320, 340. For the airframes of military aircraft, adhesives are used for primary structural bonding of fiber reinforced composite structural components. As can be seen from the SAAB 340 aircraft, two typical applications of adhesive bonding in airframes are adhesive bonded sandwich structures and adhesive bonded stiffened structures.

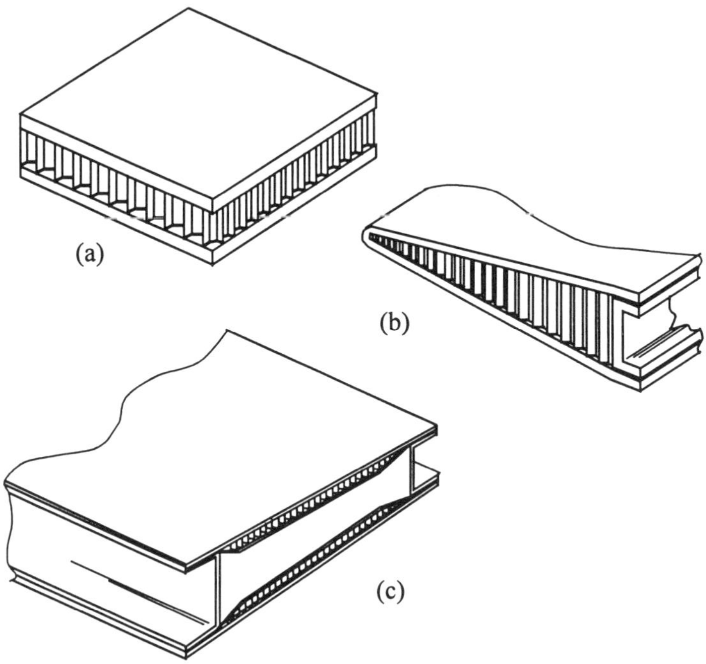

Figure 7.1 Examples of application of honeycomb structures

Adhesives have been used to manufacture honeycomb sandwich structures, in which a honeycomb core is faced with two adhesively bonded sheets of materials to produce a structure with an excellent stiffness to weight ratio. The facing materials may be metallic, such as aluminum, or non-metallic, such as carbon fiber reinforced composite laminates, or even combination of both. There is a variety of core materials, such as foam core or Nomex core. As pointed out before having the adhesive in tension is to be avoided. In honeycomb structures, adhesive fillet formed around the septum converts most of the normal force between skin and septum into shear force along the septum wall thus avoiding tension on the adhesive. Honeycomb structures have been used for flooring, sidewalls, and control surfaces, such as trailing edges and airfoils. Figure 7.1 schematically depicts typical sandwich structures of constant thickness and of variable thickness.

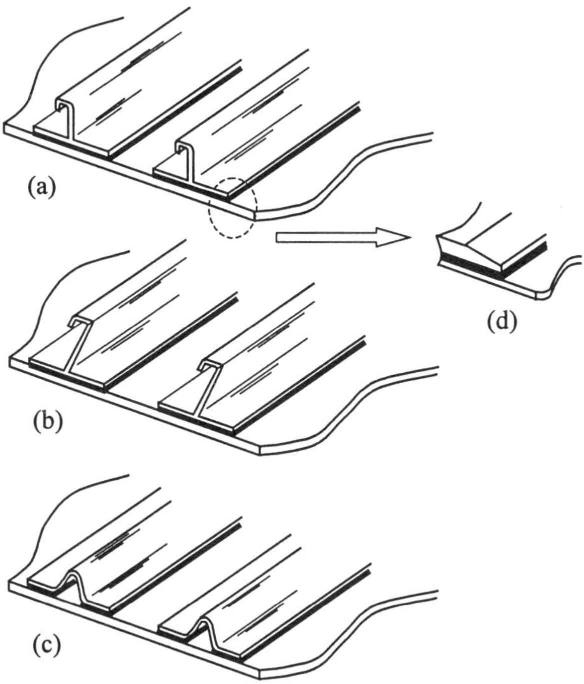

Figure 7.2 Illustrative examples of secondarily bonded stiffened panels

Adhesives are also employed to attach stiffening stringers to the relatively thin wing and fuselage skins. There are a variety of the cross-sectional shapes of the stringers, such as angle-shape, T-shape, Z-shape, hat-shape and J-shape etc. Figure 7.2(a) and 7.2(c) depict the J-shape and circular hat-shape stiffeners. The thin skin and the stiffening stringers can be either metallic or fiber reinforced composite materials, or a combination of both. The manufacturing process is often referred to as secondary bonding as opposed to co-curing when both skin and stringers are fiber reinforced composite materials. A properly designed stiffened panel has many advantages, for example; vastly increased buckling load, significant post-buckling stiffness and a failure load several times the initial buckling load can all be achieved.

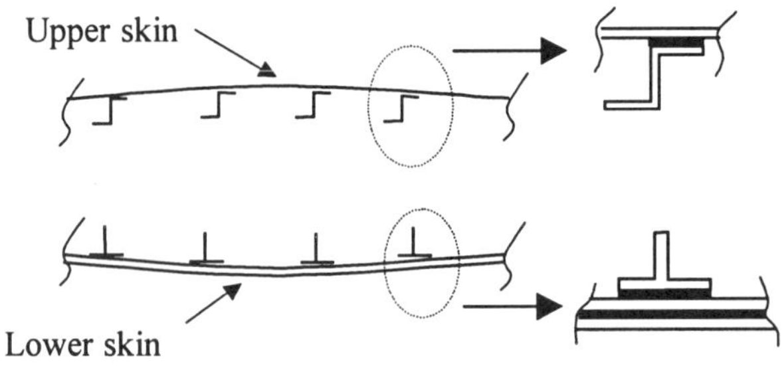

(a) Chordwise section near root

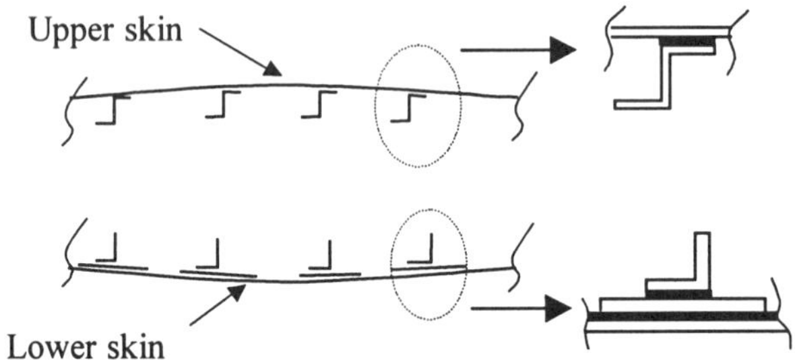

(b) Chordwise section near tip

Figure 7.3 Schematics of chord-wise section near root and tip used in SAAB 340 aircraft.

Figure 7.3 depicts schematically chordwise sections near the root and the tip of the main wing box of the SAAB 340 aircraft which was manufactured using metal-to-metal adhesive bonding and the advanced tooling developed in the PABST program. The lower skin panels consist of a skin, a second sheet comprising a continuous doubler and spanwise T-stiffeners (see Figure 7.3a) throughout the span of the fuel tanks and a waffle doubler and spanwise angle-shape stiffeners (see Figure 7.3b) at the rib and stiffener locations outboard. The upper skin panel consists of only a single sheet stiffened by Z-section stiffeners as shown in Figure 7.3. After bonding, the two panels are mechanically fastened to a pre-assembled substructure of the front and rear spars and ribs.

Separation of stringers from the skin is one of the major failure mechanisms in such structures. This occurs due to peel or normal stress concentrations developed in the adhesive near edges of adhesive layer. These stress concentrations in the bonds for undamaged structures can be alleviated by careful detail design, employing various techniques such as, tapering and fingered doubles. Figure 7.2 (b) shows a tapered stringer at its run-out, and Figure 7.2(d) depicts the tapered flange in the stringer blown up from Figure 7.2(a). As shown in previous chapters, thinning the edges of stringer flanges leads to a more flexible flange, which in turn can significantly reduce the high peel stress. Flange edge tapering of composite stringers can be achieved by dropping-off plies. This approach has been widely used in manufacturing composite airframes. Another technique currently under development is to apply transverse stitching along the flange edges of the stringers to counteract the high peel stress with an internal clamping force. This force is derived from the pretension in the stitches. The stitching also alleviates the failure of the adhesive by having much of the load diffusing into the stitches, thereby reducing adhesive peel and, in the case of angle stitching, shear stresses.

Figure 7.4 schematically shows the F-18 wing to fuselage attachment. The upper and lower skins of the graphite composite wing are bonded directly to the titanium root fitting which is part of the fuselage, using FM300K adhesive. The joint is of the step-lap joint configuration for both upper and lower skins. Compared to the FM400 adhesive, the FM300K adhesive was selected due to its overall fatigue performance and residual ultimate strength after partial fatigue exposure.

Adhesives have been used to join the various sections of helicopter rotor blades at Westland helicopters, which are complex structures, highly stressed and of limited fatigue life. A typical rotor blade may have the following sections: a stainless steel leading edge, a honeycomb trailing edge, fiber reinforced plastic skins and extruded aluminum spars. These various sections of different materials are joined together wholly by adhesive bonding to sustain a minimum static tensile strength of 7 MPa plus a further 2.1±2.1 MPa dynamic loading.

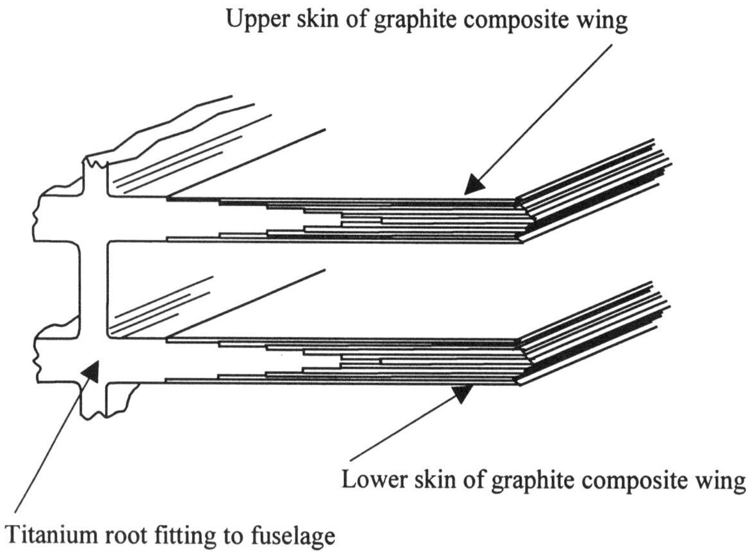

Figure 7.4 F-18 wing to fuselage attachment in which graphite composite skins are bonded to the titanium root fitting using FM 300K adhesive

Adhesive bonded composite repair represents one of the major methodologies in attempting to overcome problems associated with aging aircraft (Baker and Jones, 1988). When a crack has developed in a metallic skin or component, a fiber reinforced composite patch can be adhesively bonded to the cracked area from either side or both sides of the skin. An adequately designed and properly bonded patch can change the load path and significantly reduce the stress intensity factors at the crack tip, which in turn terminates crack growth and restores, partially or totally, the designed stiffness, strength and service life of the structure.

While composite design opportunities continue to be explored, the manufacturing cost of composite structures becomes the biggest obstacle to their widespread use. To enable affordable composite airframe structures, adhesive bonding may hold the key for the designer to fully exploit the benefits of composite materials together with fiber placement, textile structures and low cost tooling techniques.

7.3 MARINE

Applications of adhesives in the ship and boat building industries also has a long history, from wooden boat to glass reinforced plastic (GRP) ships. Many GRP hulled ships, both naval and civilian, now employ a significant amount of adhesive for laminating, stiffening, fabricating sandwich panels, and for bonding various attachments as well as gap filling.

In many vessels, non-structural sandwich and insulating panels are fabricated using a variety of combinations of bonded skin and core materials, and are employed in a large-scale in the fitting-out of the vessel. In fact, one of the most important marine applications of adhesives is in the form of paint corrosion prevention system for hulls.

There are also a number of structural applications of adhesives in vessels (Pei and Shenoi, 1996). Joints become necessary in the structure of ships and boats made from fiber reinforced plastic materials because large and complex structures can not be formed in one process. Structural members can be connected by bonding or laminating. Such connections represent a zone of potential weakness in a large fiber reinforced plastic structure. In marine structures it is typical for the two joining members are orthogonal to each other, which is referred to as out-of-plane joints compared to the in-plane lap joints. Examples of out-of-plane joints include frame-to-shell, bulkhead-to-shell, stiffener ending, stiffener intersection, deck edge in tee form and deck edge in knee form. These joints can be classified into two groups as schematically shown in Figure 7.5. Figure 7.5(a) and (b) depict the cross-sectional view of two types of tee joints, which represent connections between two orthogonally placed plate panels, such as bulkhead-to-deck, shell-to-bulkhead, deck-edge in tee form. In these joints, the vertical and horizontal plating panels are preformed either in single sheet or sandwich form and the joint is then fabricated with access to both sides of the vertical panel either using in-situ laminating or secondary bonding. The design variables of a tee joint can include the shape, material makeup of plies and number of plies of the boundary angles (see Figure 7.5), the gap size between the two panels, the types of resin or adhesive and the shape of the bottom edge of the vertical panel. Figure 7.5(c) depicts the cross-sectional view of a typical top-hat stiffener, which is representative of frame-to-shell connections. The top-hat section can be moulded in situ on to the cured shell or preformed and then attached to the cured shell by secondary bonding. Another alternative is to preform the U-section and then connect it in place using adhesive bonding and two boundary angles.

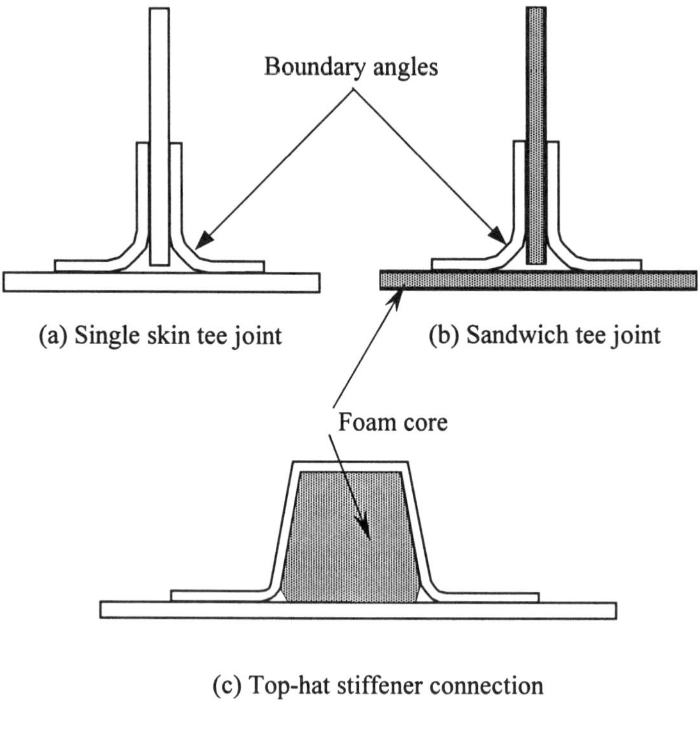

Figure 7.5 Cross-sectional views of typical out-of-plane joints used in marine structures. (a) single skin tee joint, (b) sandwich tee joint, and (c) top-hat stiffener joint

The weakness of tee joints and top-hat stiffeners is primarily due to the lack of reinforcement across the connected surfaces, the occurrence of stress concentrations associated with joint geometry and production considerations. For the two types of tee joints, namely, single skin tee joints and sandwich tee joints, performance is often evaluated using the following three types of testing methods: (a) tensile test, (b) compressive test, and (c) three-point bending test. These three tests are schematically shown in Figure 7.6. For the top-hat stiffener connections, it is vitally important to ensure that a good bond is being achieved between the top-hat stiffener and the plating panels. The performance of these joints has been studied by using the three test methods sketched in Figure 7.7.

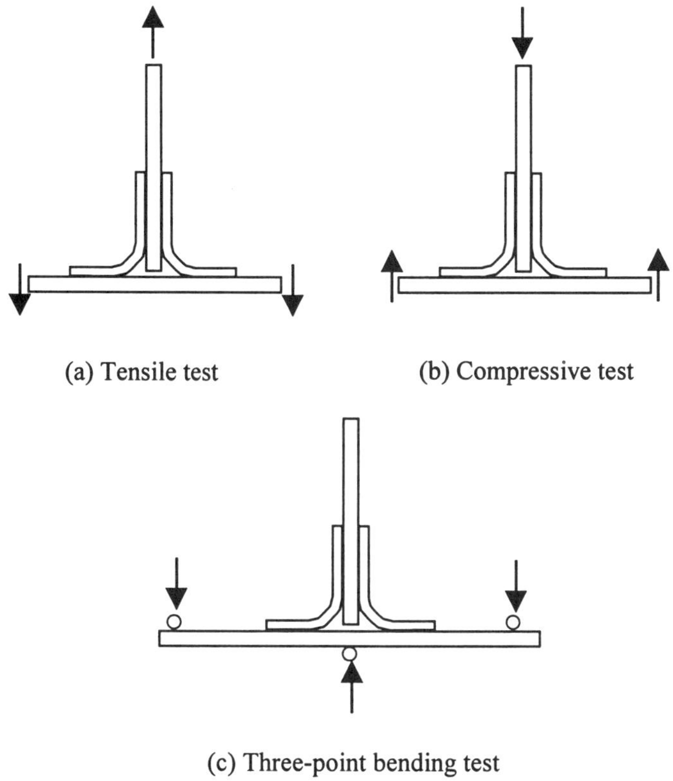

Figure 7.6 Three test methods for evaluating tee joint performance. (a) tensile test, (b) compressive test, and (c) three-point bending test

Recently, tests have been performed to understand the effect of various design variables on the static strength and failure modes of both tee and top-hat stiffener joints using the test methods shown in Figures 7.6 and 7.7 (see, e.g., Shenoi and Violette, 1990; Shenoi and Hawkins, 1992; Theotokoglou and Moan, 1996; Pei and Shenoi, 1996; Rispler et al, 1997a,b). In addition to static strength, response under impulsive loading, creep and fatigue characteristics have also been investigated.

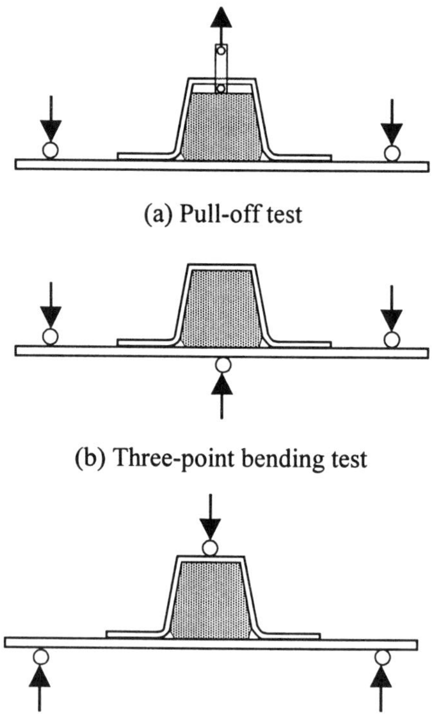

(a) Pull-off test

(b) Three-point bending test

(c) Reverse three-point bending test

Figure 7.7 Three test methods for evaluating the performance of the top-hat stiffener joints. (a) pull-off test method, (b) three point bending, and (c) reverse three-point bending test

The first glass fiber reinforced plastic (GFRP) boat was made in America in 1944. Half a century of design, production and service experiences has resulted in a number of design codes and rules for the fabrication of GFRP marine structures. For example, the Lloyd's Register of Shipping rules

(1983), the American Bureau of Shipping rules (1978), and Det Norske Veritas rules (1985) are well-known design rules for GFRP marine structures. The standards cover the methods of joining different structural members and design of structural intersections including tolerances, such as single skin tee joints, sandwich tee joints and top-hat stiffener joints. For example, the American Bureau of Shipping rules specifies the thickness of each boundary angle as shown in Figure 7.5(a) and 7.5(b). The thickness of each angle shall be less than one-half the thickness of the thinner of the two laminated panels being joined for a laminated single skin tee joint, and shall be the thickness of one skin of the thinner of the sandwich panels for a sandwich tee joint.

Similar to the aerospace industry, bonded repair of metal superstructures represents another important structural application of adhesives. Fatigue cracking may develop in metallic superstructures, and it is sometime difficult to stop crack growth simply by drilling out the crack tips. New technology has been developed using adhesive by bonding carbon fiber reinforced laminates over the crack affected area. For example, two-part cold-curing epoxy adhesives have been employed to bond carbon fiber reinforced laminate to stop cracking developed in Type 21 Frigates (Mays and Hutchinson, 1992).

7.4 GROUND TRANSPORTATION

Adhesives are being widely used in the car, truck, train and even bicycle manufacturing industries. Adhesive bonding of stiffeners to bonnet and boot lids represents one of the earlier metal applications in car body manufacturing. The stiffeners are needed to overcome flutter and rattle problems at high speed. Two typical candidate applications for adhesive bonding are flange seams and clinched seams in door and roof assemblies, as schematically shown in Figure 7.8. Figure 7.9 depicts a bicycle prototype manufactured in the Department of Aeronautical Engineering at the University of Sydney by bonding the two carbon fiber reinforced plastic thin shells together.

Other applications on vehicles include attaching trim, thread locking, gasketing in the engine, headlamps, radiators and brake linings. In general, applications of adhesive in car making industry can be divided into the usual two groups, non-structural applications and structural applications. There are three general areas of non-structural applications, (a) all interior trim, such as the roofliner assembly and attachment, side-door covering, dashboard assembly and other decorative coverings; (b) exterior-trim attachments, such as front and rear tail-light assemblies, side-mirror assemblies; and (c) all under-the-hood applications, e.g. the form-in-place sealant. Structural applications can be found in the following areas: (a) bonded brakes and other clutch assemblies; (b) windshield and back-light attachment, first introduced in 1979 and now widely used in most car models; (c) exterior closure-panel attachment; (d) hem-flanges for doors, hoods and deck lids; and (e) adhesive patching.

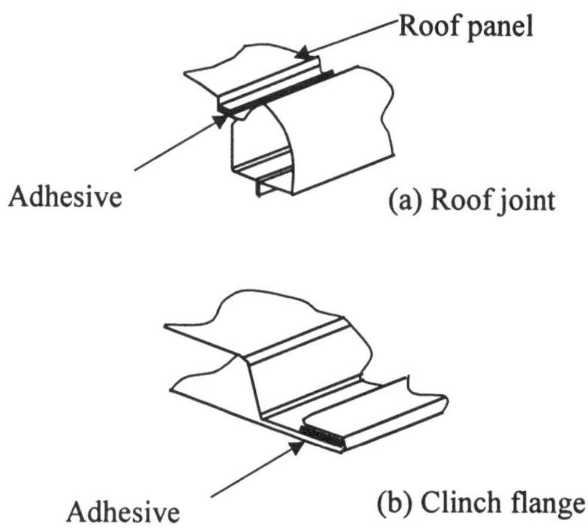

Figure 7.8 Schematics of a roof joint and a clinch flange joint

Lightweight materials, such as aluminum alloys, plastics and fiber reinforced composite materials, are also used in car construction industry to help achieve improved fuel economy, low exhaust emissions and higher performance. Aluminum alloys can not be readily spot-welded, therefore adhesive bonding is employed as the joining method. One of the major advantages of adhesive bonding is its capability of joining dissimilar materials.

One innovative application of adhesive bonding is that in the Peugeot rally car where a carbon fiber reinforced plastic drive shaft is bonded into the metal end-fittings with a toughened-epoxy polymer. The torque from the gearbox passes through the fittings to the drive shaft then again through the adhesive to the differential. The shear stress in the adhesive layer transmits the high torque generated in this high-powered rally car. This carefully shaped co-axial joint design has been used successfully in many other rally cars. Another example of lightweight vehicle comprising all three materials is the British Leyland 'Energy Conservation Vehicle Mark 3' (BL ECV3), in which very significant use of adhesive bonding was implemented. The structural integrity of the car is derived from the aluminum alloy monocoque body which was wholly joined together using a toughened epoxy resin.

Figure 7.9 Bicycle with bonded optimal designed fiber reinforced frame

In suburban and inter-city coaches in the UK, adhesives have been used to fabricate secondary structural components. For example, sandwich doors made by bonding aluminum or GRP skins onto honeycomb materials, aluminum ceiling panels or steel side body panels stiffened by bonding hat-section stiffeners.

Automatic and robotic application of adhesives is a fast growing field, and it offers many advantages to general engineering industry. In car production line, robotic automatic adhesive dispensing equipment has been installed for component assembly. The use of this type of equipment has enabled a consistent application quality to be achieved for mass production with careful control and monitoring of the process and materials used.

Weld bonding is probably the newest and most exciting application of adhesive in car industry. This process uses both welding and adhesive bonding to achieve improved structural integrity and stiffness not available to either spot welding or adhesive bonding alone, and can be used to join structural frames of a vehicle. The 1997 Ford Expedition uses structural adhesive weld bonding around the lift-gate opening and other areas to improve the stiffness and durability of the steel-body structure and the noise, vibration and harshness characteristics of the vehicle when in use. Some other examples include the bonding of the side aperture, the inner to outer rocker panels, various cross bracing and the rear-shelf panel. To increase the applications of weld bonding, it is important to understand the fundamentals of the various design parameters and adhesive properties in the weld bonding process.

7.5 CIVIL

In the building industry the majority of adhesive bonding applications are concerned with fastening decorative finishing materials to the insides of buildings. For instance, ceiling tiles, ceramic tiles and mosaics bonded to floors and walls. There are also applications for adhesives in external situations that have greater structural demands. One of the major constraints is the specifications and applications of materials under on-site conditions, which can be significantly different from in-factory conditions and are very difficult to control.

Some civil structural components may deteriorate after many years in-service. In many cases, it is preferred to retrofit the defective components than to rebuild or replace. Maintenance of aging civil structures is becoming a fast growing field. Once again, adhesives have been selected to repair and strengthen defected civil structures, such as concrete structures, masonry structures. To repair a cracked concrete structure, it is necessary to identify whether the crack is inactive or live. For an inactive crack, resin injection technique has been used to restore structural integrity by bridging the two crack surfaces and durability by sealing the cracks. For a live crack, a structural repair must be performed to prevent further movements of the two crack surfaces followed by resin injection for sealing.

It is sometimes necessary to increase the reinforcement of existing concrete structures due to, for example, design deficiencies, construction errors, or even due to amendments in the design code. Adhesives are employed to join the fresh and the older hardened concrete following the concrete overlay process. Adhesive are also used to bond a mild steel plate to the external surface of the weak or defected concrete component in order to control further development of the damage. Adhesives have been employed to build composite steel-concrete constructions and segmental concrete constructions.

Resin bonded methods have been developed in early 1970s to repair timber structures. Resin bonded repairs to timber structures have been used in the following five circumstances: beam end repair, trussed rafter and foot repair, column repair, fissure repairs and upgrading beams.

References

Adams, R.D. (1989) Strength predictions for lap joints, especially with composite adherends: a review. *J. of Adhesion*, **30** 219-242.

Adams, R.D., Atkins, R.W., Harris, J.A. and Kinloch, A.J. (1986) Stress analysis and failure properties of carbon-fibre-reinforced-plastic/steel double-lap joints. *J. of Adhesion*, **20** 29-53.

Adams, R.D. and Coppendale J. (1977) *Adhesion 1*. (Allen, K.W., ed.), Applied Science Publishers, London, 1-17.

Adams, R.D. and Coppendale J. (1979) The stress-strain behavior of axially-loaded butt joints. *J. Adhesion*, **10** 49-92.

Adams, R.D., Comyn, J. and Wake W.C. (1997) *Structural Adhesive Joints in Engineering*, 2nd Edition, Chapman & Hall, London.

Adams, R.D. and Harris, J.A. (1987) The Influence of local geometry on the strength of adhesive joints. *Int. J. of Adhesion and Adhesives*, **7** (2) 69-80.

Adams, R. D. and Mallick, V. (1992) A method for the stress analysis of lap joints. *J. of Adhesion*, **38** 199-217.

Adams, A.D. and Peppiatt, N.A. (1974) Stress analysis of adhesive-bonded lap joints. *J. of Strain Analysis*, **9** (3) 185-196.

Adams, A.D. and Peppiatt, N.A. (1977) Stress analysis of adhesive bonded tubular lap joints. *J. of Adhesions*, **9** (1) 1-18.

Adams, R.D. and Wake, W.C. (1984) *Structural Adhesive Joints in Engineering*, Elsevier Applied Science Publishers, London.

Allman, D.J. (1977) A theory for elastic stresses in adhesive bonded joints. *Quart. J. Mech. Appl. Math.*, **30** 415-436.

Althof, W. and Brockmann, W. (1976) New test methods for the prediction of environmental resistance of adhesive bonded joints. *Bicentennial of Materials Progress*, **21** 581-591.

Althof, W. (1979) Bonded joints and preparation for bonding. AGARD lecture series No. 102, March.

Alwar, R.S. and Nagaraja, Y.R. (1976) Viscoelastic analysis of an adhesive tubular joints. *J. of Adhesion*, **8** 79-92.

American Bureau of Shipping (1978) *Rules for building and classing reinforced plastic vessels*, New York, USA

Anderson, G.P., Brinton, S.H., Ninow, K.J. and DeVries, K.L. (1988) A fracture mechanics approach to predicting bond strength. in *Advances in Adhesively Bonded Joints*, ASME New York, MD **6**, 93-101.

Anderson, G.L., Robertson, R.C., Peterson, B. L. and Dillard, D.A. (1994) Embedded piezoelectric sensors to measure peel stresses in adhesive joints. *Experimental Mechanics*, **34** (3) 194-201.

Anderson, T.L. (1995) *Fracture Mechanics: Fundamentals and Applications*. CRC Press, Boca Raton, Florida, USA.

Anonymous (1986) Boeing Advanced Composites Handbook, Seattle, Washington, USA, June.

Anonymous (1989) ICI Fiberite Materials Handbook, Tempe, Arizona, USA, March.

Atluri, S.N. (1997) *Structural Integrity and Durability*. Tech Science Press. Forsyth, Georgia, USA.

Baker, A. A. and Jones, R. (1988). *Bonded repair of aircraft structures.* Martinus Nijhoff, Dordrecht.

Beaumont, P.W. and Schultz, J.M. (1990) Macroscopic fracture theories in Failure Analysis of Composite Materials, *Delaware Composites Design Encyclopedia*, edited by Beaumont, P.W., Schultz, J.M. and Friedrich, K., Technomic Pub. Co., Lancaster, USA, **4** 83-115.

Bendsoe, M.P. (1995) Optimization of Structural Topology, Shape, and Material, Springer, Heidelberg.

Bigwood, A. and Crocombe, A.D. (1989) Elastic analysis and engineering design formulae for bonded joints. *Int. J. of Adhesion and Adhesives*, **9** (4) 229-242.

Bradley, W. L. and Cohen, R. N. (1985) Matrix deformation and fracture in graphite-reinforced epoxies. *ASTM STP*, **876** 389-410.

Brussat, T.R., Chiu, S.T. and Mostovoy, S. (1977) Fracture mechanics for structural adhesive bonds, *AFNL-TR-77-163*, Air Force Materials Laboratory, Wright-Patterson AFB, Ohio, USA.

Carpenter, W.C. (1989) Goland and Reissner were correct. *J. of Strain Analysis*, **24** (3).

Carpenter, W.C. (1991) A comparison of numerous lap joint theories for adhesively bonded joints. *J. of Adhesion*, **35** (1) 55-73.

Carpenter, W. C., Read, D. T. & Dodds, R. H. Jr (1986) Comparison of several path independent integrals including plasticity effects. *Int. J. of Fracture,* **31** 303-323.

Chai, H (1987) A note or crack trajectory in an elastic strip bounded by rigid substrates. *Int. J. Fracture*, **32** 211-213.

Chai, H. (1993a) Observation of deformation and damage at the tip of cracks in adhesive bonds loaded in shear and assessment of a criterion for fracture. *Int. J. Fracture*, **60** (4) 311-326.

Chai, H. (1993b) Deformation and failure of adhesive bonds under shear loading. *J. of Materials Science*, **28** (18) 4944-4956.

Chang, M.D., Devries, K.L. and Williams, M.L. (1972) The effects of plasticity in adhesive fracture. *J. Adhesion,* **4** 221.

Chen, D. and Cheng, S. (1983) An analysis of adhesive-bonded single-lap joints. *ASME J. of Applied Mechnics*, **50** (March) 109-115.

Cherry, B.W. and Harrison, N.L. (1970) The optimal profile for lap joint. *J. Adhesion,* **2** 125-128.

Chon, C.T. (1982) Analysis of tubular joint in torsion. *J. of Composite Materials*, **16** 268-284.

Clark, J.D. and McGregor, I.J., (1993) Ultimate tensile stress over a zone: a new failure criterion for adhesive joints. *J. of Adhesion*, **42** (4) 227-245.

Cornell, R. W., (1953) Determination of stresses in cemented lap joints. *ASME J. of Applied Mechanics.*, **20** 355-364.

Crocombe, A.D., (1989) Global yielding as a failure criteria for bonded joints. *Int. J. of Adhesion and Adhesives*, **9** (3) 145-153.

Crocombe, A.D. and Adams, R.D. (1981) Influence of the spew fillet and other parameters on the stress distribution in the single lap joint. *J. of Adhesion*, **13** 141-155.

Crocombe, A.D. and Adams, R.D. (1982) An elasto-plastic investigation of the peel test. *J. of Adhesion*, **13** 241-267.

Crocombe, A.D., Bigwood, D.A. and Richardson G. (1990) Analysing structural adhesive joints for failure. *Int. J Adhesion and Adhesives*, **10** (3) 167-178.

Curley, A. J., Jethwa, J.K. and Kinloch, A.J. (1998) The fatigue and durability behaviour of automotive adhesives. Part III: Predicting the service life. *J. Adhesion*, **66** 39-59.

Czarnocki, P. and Piekarski, K. (1986) Fracture strength of an adhesive-bonded joint. *Int. J. of Adhesion and Adhesives*, **6** (2) 93-95.

Delale, F. (1984) Stress singularities in bonded anisotropic materials. *Int. J. of Solids and Structures*, **20** (1) 31-40.

Det Norske Veritas (1985) *Rules for classification of high speed light craft.* Norway.

Dillard, D.A., Anderson, G.L. and Davis, Jr., D.D. (1989) A preliminary study of the use of Kynar piezoelectric film to measure peel stresses in adhesive joints. *J. of Adhesion*, **29** 235.

Ding, Y. (1986) Shape optimisation of structures: a literature survey. *Comput. Struct.* **24** (6) 985-1004.

Dolev, G. and Ishai, O. (1981) Mechanical characterization of adhesive layer 'in-situ' and as 'bulk' material. *J of Adhesion,* **12** 289-294.

Dowling, N.E. and Begley, J.A. (1976) Fatigue Crack Growth During Gross Plasticity and the J Integral. *ASTM STP* **590** 82-103.

Dransfield, K., Baillie, C. and Mai, Y.W. (1994) Improving the delamination resistance of CFRP by stitching - A review. *Composite Science and Technology*, **50** (3) 305-317.

Edde, F. and Verreman, Y. (1992) On the fracture parameters in a clamped cracked lap shear adhesive joint. *Int. J. of Adhesion and Adhesives,* **12** 43-48.

Edlund, U. and Klarbring, A. (1993) A coupled elastic-plastic damage model for rubber-modified epoxy adhesives. *Int. J. of Solids and Structures*, **30** (19) 2693-2708.

Edlund, U. (1994) Surface adhesive joint description with coupled elastic-plastic damage behaviour and numerical applications. *Computer Methods in Applied Mechanics and Engineering*, **115** (3-4) 253-276.

Engineering Sciences Data Unit (1979) Inelastic shear stresses and strains in the adhesives bonding lap joints loaded in tension or shear. *Item Number 79016*, London.

Engineering Sciences Data Unit (1980) Elastic adhesive stresses in multistep lap joints loaded in tension, *Item Number 80039*, London.

Engineering Sciences Data Unit (1981) Guide to the use of data items in the design of bonded joints, *Item Number 81022*, London.

Erdogan, F. and Ratwani, M. (1971) Stress distribution in bonded joints. *J. Composite Materials*, **5** 378-393.

Falzon, B., Steven, G.P. and Xie, Y.M. (1996) Shape optimization of interior cutouts in composite panels. *Struct. Optim.* **11** 43-49.

Fernlund, G., Papini, M., McCammond, D., et al. (1994) Fracture load predictions for adhesive joints. *Composite Science and Technology*, **51** (4) 587-600.

Fernlund, G. and Spelt, J.K. (1994a) Mixed-mode fracture characterization of adhesive joints. *Composites Science and Technology*, **50** (4) 441-449.

Fernlund, G. and Spelt, J.K. (1994b) Mixed mode energy release rates for adhesively bonded beam specimens. *J. of Composites Technology and Research*, **16** (3) 234-243.

Fleck, N.A., Hutchinson, J.W. and Suo, Z. (1991) Crack path selection in a brittle adhesive layer. *Int. J. of Solids and Structures*, **27** (13) 1683-1703

Gali, S., Dolev, G. and Ishai, O. (1981) An effective stress/strain concept in the mechanical characterization of structural adhesive bonding. *Int. J. Adhesion and Adhesives*, **1** (3) 135-140.

Gent, A.N. and Hamed, G.R. (1977) Peel mechanics for an elastic-plastic adherend. *J. Applied Polymer Science*, **21** 2817.

Gere, J.M and Timoshenko, S. P. (1991) *Mechanics of Materials*. 3rd ed., Chapman & Hall, London.

Goland, M. and Reissner, E. (1944) The stresses in cemeneted joints. *ASME J. of Applied Mechanics*, **7** (March) A17-27.

Goldberg, D.E. (1989*) Genetic Algorithms in Search, Optimisation and Machine Learning*, Addison-Wesley.

G+D Computing Pty. Ltd. (1995) *STRAND6 user manual*. Sydney, Australia.

Groth, H. L. (1988) Stress singularities and fracture at interface corners in bonded joints. *Int. J. of Adhesion and Adhesives*, **8** (2) 107-113.

Groth, H. L. and Nordlund, P. (1991) Shape optimization of bonded joints. *International Journal of Adhesion and Adhesives,* **11** (4) 204.

Haftka, R.T. and Grandhi, R.V. (1985) Structural shape optimisation: a survey. AIAA-85-0772, AIAA/ASME/ASCE/AHS 26th Struct. Dyn. Mat. Conf., Florida.

Haftka, R.T. and Gurdal, Z. (1992) *Elements of Structural Optimization*. 3rd edn., Kluwer Academic Publishers, Dordrecht.

Harris, J. and Adams, R. (1984) Strength prediction of bonded single lap joints by non-linear finite element methods. *Int. J. of Adhesion and Adhesives*, **4** (2) 65-78.

Hart-Smith, L. J. (1973a) *Adhesive-bonded double-lap joints*. NASA CR-112235, NASA Langley Research Centre, Hampton, Virginia, USA.

Hart-Smith, L. J. (1973b) *Adhesive-bonded single-lap joints*. NASA CR-112236, NASA Langley Research Centre, Hampton, Virginia, USA.

Hart-Smith, L. J. (1973c) *Adhesive-bonded scarf and stepped-lap joints*. NASA CR-112237, NASA Langley Research Centre, Hampton, Virginia, USA.

Hart-Smith, L.J. (1973d) *Non-Classical Adhesive-Bonded Joints in Practical Aerospace Production*. NASA CR-112238, NASA Langley Research Centre, Hampton, Virginia, USA.

Hart-Smith, L.J. (1978) Adhesive bond stresses and strains at discontinuities and cracks in bonded structures. *ASME J. Engineering Materials and Technology*, **100** 16-24.

Hart Smith, L. J. (1987) *Joining Fibre-Reinforced Plastics*. Ed. by Mathews F. L., Elsevier Applied Science, Chapter 7, 282-299.

Hart-Smith, L.J. (1988). Design and analysis of bonded repairs for metal aircraft structures. *Bonded Repair of Aircraft Structures*, Baker, A. A. and Jones, R. eds, Martinus Nijhoff, Dordrecht, Chapter 3.

Hart-Smith, L.J. and Strinberg, G. (1997) Developments in adhesively bonding the wings of the SAAB 340 and 2000 aircraft. *Proc. Instn. Mech. Engrs. J. Aerospace Eng.* Part G, **211** 133-156.

Hashemi, S., Kinloch, A. J. & Williams, G. (1991) Mixed mode fracture in fiber-polymer composite laminates. *ASTM STP.*, **110** 143-62.

Hattori, T. (1991) A stress-singularity-parameter approach for evaluating the adhesive strength of single-lap joints. *JSME Int. J. Series I*, **34** (3) 326-331.

Hein, V.L. and Erdogan, F. (1971) Stress singularities in a two-material wedge. *Int. J. of Fracture Mechanics*, **7** (3) 317-330.

Hoskin, B.C. and Baker, A.A. (1986) *Joining Advanced Fiber Composites. Composite Materials for Aircraft Structures*, AIAA Education Series, Chapter 8, 115-139.

Hollaway, L., Romhi, A. and Gunn, M. (1990) Optimisation of adhesive bonded composite tubular sections. *Composite Structures*, **16** 125-170.

Hutchinson, J.W. and Suo, Z. (1992) Mixed mode cracking in layered materials. *Advances in Applied Mechanics*, **29** 63-191.

Ikegami, K., Sugibayashi, T. and Matsui, K. (1995) Strength of cylinder joint adhesively bonded by coupling. *Archive of Applied Mechanics*, **65** (1) 44-53.

Irvin, G.R. (1956) Onset of fast crack propagation in high strength steel and aluminium alloys. *Sagamore Research Conference Proceedings*, **2** 289-305.

Irvin, G.R. (1958) Fracture. *Handbook der Physik*, **6** 551.

Jain, L.K, Dransfield, K, Mai, Y-W and Baillie, C. (1995) Improvement of interlaminar properties in advanced fibre composites with through-thickness reinforcement. CRC-AS annual report, Melbourne, Australia.

Jain, L.K., Leong, K.H., Mai, Y.W. and Tong, L. (1998) Effect of through-thickness stitching on the fatigue life of composite single-lap joints. *Applied Composite Materials*. **5** (6) 399-409

Jeandrau, J.P. (1986) Intrinsic mechanical characterization of structural adhesives. *Int. J. Adhesion and Adhesives,* **6** (4) 229-231.

Jeandrau, J.P. (1991) Analysis and design data for adhesively bonded joints. *Int. J. Adhesion and Adhesives,* **11** (2) 71-79.

Jethwa, J.K. and Kinloch, A. J. (1997) The fatigue and durability behaviour of automotive adhesives. Part I: fracture mechanics tests. *J. Adhesion,* **61** 71-95.

John, S.J. (1994) Predicting the strength of bonded carbon fibre/epoxy composite joints. *Composites Bonding*, ASTM STP 1227, edited by Damico, D. J., Wilkinson, T. L. and Niks, S. L. F., ASTM, Philadelphia, USA, 45-59.

John, S.J., Kinloch, A.J. and Matthews, F.L. (1991) Measuring and predicting the durability of bonded carbon fibre/epoxy composite joints. *Composites*, **22** (2) 121-126.

Johnson, W.S. (1987) Stress analysis of the cracked-lap-shear specimen: an ASTM round robin. *J. of Testing and Evaluation*, **6** 303-324.

Jones, R., Chiu, W. K. and Paul, J. (1993). Designing for damage tolerant bonded joints. *Composite Structures*, **25** 201-207.

Kairouz, K.C., and Matthews, F.L. (1993) Strength and failure modes of bonded single lap joints between cross-ply adherends. *Composites*, **24** (6) 475-484.

Kan, H.P. and Ratwani, M.M. (1983) Stress analysis of stepped-lap joints with bondline flaws. *J. of Aircraft*, **20** (10) 848-852.

Kassapoglou, C. (1993) Calculation of stresses at skin-stiffener interfaces of composite stiffened panels under shear loads. *Int. J. of Solids and Structures*, **30** (11) 1491-1501.

Kassapoglou, C. (1994) Stress determination at skin-stiffener interfaces of composite stiffened panels under generalised loading. *J. of Reinforced Plastics and Composites*, **13** (6) 555-572.

Kassapoglou, C. and Adelmann, J.C. (1991) KGR-1 Thick adherend specimen evaluation for determination of adhesive mechanical properties. *23rd Int. SAMPE Technical Conference*, Ed. Carri, R.L., Poveromo, L.M. and Gauland, J., Vol. **23** 162-176.

Kim, K.-S. And Aravas, N. (1988) Elastoplastic analysis of the peel test. *Int. J. Solids Structures*, **24** 417-435.

Kinloch, A.J. (1987) *Adhesion and adhesives: science and technology*, Chapman and Hall, London.

Kinloch, A.J. (1996) Adhesives in Engineering, 84th Thomas Hawksley Memorial Lecture, Proceedings of the Institution of Mechanical Engineers, December 1996.

Knauss, W.G. and Losi, G.U. (1993) Crack propagation in a nonlinearly viscoelastic solid with relevance to adhesive bond failure. *J. of Applied Mechanics*, **60** 793-801.

Krieger, R.B., Jr. (1988) Stress analysis concepts for adhesive bonding of aircraft primary structures. *Adhesively Bonded Joints: Testing, Analysis and Design, ASTM STP,* **981**, W. S. Johnson, Ed., American Society for Testing and Materials, Philadelphia, USA, 264-275.

Kristensen, E.S. and Madsen, N.F. (1972) On the optimum shape of fillets in plates subject to multiple in-plane loading cases, *Int. J. Num. Meth. Engng.* **10** 1007-1019.

Lai, Y.-H., Rakestraw, M.D. and Dillard, D.A. (1996) The cracked lap shear specimen revisited - a closed form solution. *Int. J. of Solids and Structures*. **33** (12) 1725-1743.

Laksimi, A., Gong, XL and Benzeggagh, M.L. (1994) Analysis of damage mechanisms in centrally notched glass-fiber epoxy plate, Composite Science and Technology, **52** (1) 85-91.

Lamba, H.S. (1975) The J-Integral Applied to Cyclic Loading. *Engineering Fracture Mechanics,* **7** 693-703.

Lambert, Y., Sailard, P., and Bathias, C. (1988) Application of the J Concept to Fatigue Crack Growth in Large-Scale yielding. *ASTM STP,* **969** 318-329.

Lee, C. and Liu, D. (1990) Tensile strength of stitching joint in woven glass fabrics. *J. of Engineering Materials and Technology*, **112** 125-130.

Lee, S.J., and Lee, D.G. (1992) Development of a failure model for the adhesively bonded tubular single lap joint. *J. of Adhesion*, **40** (1) 1-14.

Lerchenthal, C. H. (1964) *Design of Bonded Lap Joints for Optimum Stress Transfer, Part 1. The Symmetrical (or double) Lap Joint*. Department of Mechanics, Technion-Israel Institute of Technology, Haifa.

Lin, C. and Liechti, K.M. (1987) Similarity concepts in the fatigue fracture of adhesively bonded joints. *J. Adhesion*, **21** 1-24.

Liu, D. (1990) Delamination resistance in stitched and unstitched composite planes subjected to composite loading. *J. of Reinforced Plastics and Composites*, **9** 59-69.

Lloyd's Register of Shipping (1983) *Rules for yachts and small craft*. London, UK.

Long, R.S. (1991) Static strength of adhesively bonded ARALL-1 joints. *J. of Composite Materials*, **25** (April) 391-415.

Lubkin, J.L. (1957) A theory of elastic scarf joints. *ASME J. of Applied Mechanics*, **24** (June) 255-260.

Lubkin, J.L. and Reissner, E. (1956) Stress distribution and design data for adhesive lap joints between circular tubes. *Transactions of ASME*, **78** 1214-1221.

MacNeal-Schwendler Corporation (1994) *MSC/NASTRAN V68 User's Manual*.

Martin, R.H. and Murri, G.B. (1990) Characterization of mode I and mode II delamination growth and thresholds in AS4/PEEK composites, *Composite Materials: Testing and Design* (9th Volume), *ASTM STP,* 1059, S.P. Garbo, Ed., American Society for Testing and Materials, Philadelphia, USA, 251-270.

Matsui, K. (1990) Size-effects on average ultimate shear stresses of adhesive-bonded rectangular or tubular lap joints under tension-shear. *J. of Adhesion*, **10** (2) 81-89.

Mattheck, C. and Burkhardt, S. (1990) A New Method of Structural Shape Optimisation Based on Biological Growth, *Int. J. Fatigue*, 12(3), pp185-190.

Matthews, F.L., Kilty, P.F. and Godwin, E.W. (1982) A review in the strength of joints in fibre reinforced plastics. *Composites*, **13** (1) 29-37.

Mays, G.C. and Hutchinson, A.R. (1992) *Adhesives in civil engineering,* Cambridge University Press, Cambridge.

McClintock, F. A. (1958) Ductile fracture instability in shear. *J. of Applied Mechanics*.**25** 582-587.

Mignery, L.A., Sun, C.T. and Tan, T.M. (1985) The use of stitching to suppress delamination in laminated composites. *ASTM STP,* **876** 371-385.

Moore, G. J. (1994) "*MSC/NASTRAN Design Sensitivity and Optimization - User's Guide*", The MacNeal-Schwendler Corporation

Morales, A. (1990) Structural stitching of textile preforms. *Proceedings of the 22nd Int. SAMPE Technical Conference*, 6-8 Nov. 1990, 1217-1230.

Mori, K. and Sugibayashi, T. (1990) Predicting the strength of stepped-lap bonded joint with adhesive resin under tensile shear load. *JSME Int. J. Series I*, **33** (3) 349-355.

Mostovoy, S. and Ripling, E.J. (1975) Flaw tolerance of a number of commercial and experimental adhesives. *Adhesion Science and Technology*, **9B**, 513-562.

Nahas, M.N. (1986) Survey of failure and post-failure theories of laminated fibre-reinforced composites. *J. of Composites Technology and Research*, **8** (4) 138-153.

O'Brien, T.K. (1982) Characterization of delamination onset and growth in composite laminate. *Damage in Composite Materials*, *ASTM STP*, **775**, 140-167.

Ogo, Y. (1987) The effect of stitching on in-plane and interlaminar properties of carbon/epoxy fabric laminates. MSc. Thesis, University of Delaware, Newark, Delaware, USA.

Ojalvo, I. U. (1985) Optimization of bonded joints. *AIAA Journal*, **23** (10) 1578-1582.

Ojalvo, I. and Eidinoff, H. (1978) Bond thickness effects upon stresses in single-lap adhesive joints. *AIAA Journal*, **16** 204-211.

Olhoff,N, Bendsoe, M.P. and Rozvany, G. (1991) On CAD-integrated structural topology and design optimisation, *Comp. Meth. Appl. Mech. Engng*, **89** 259-279.

Oplinger, D.W. (1991) *A layered beam theory for single lap joints*. Army Materials Technology Laboratory Report MTL TR91-23.

Owen, D.R.J. and Hinton, E. (1980) *Finite Elements in Plasticity: Theory and Practice*, Chapter 7, Pineridge Press Limited, Swansea, UK.

Papini, M., Fernlund, G. and Spelt, J.K. (1994) Effect of crack-growth mechanism on the prediction of fracture load of adhesive joints. *Composites Science and Technology*, **52** (4) 561-570.

Paris, P.C. and Erdogan, F. (1960) A Critical Analysis of Crack Propagation Laws. *J. of Basic Engineering*, **85** 528-534.

Paris, P.C., Gomez, M.P., and Anderson, W.P. (1961) A Rational Analytic Theory of Fatigue. The Trend in Engineering, **13** 9-14.

Pedersen, P., Tobiesen, L. and Jensen, S.H. (1992) Shapes of Orthotropic Plates for minimum Energy Concentration, *Mech. Struct and Mach*, **20** (4) 499-514.

Pei, J.H. and Shenoi, R.A. (1996) Examination of key aspects defining the performance characteristics of out-of-plane joints in FRP marine structures. *Composites: Part A Applied Science & Manufacturing*, **27A** (2) 89-103.

Pickett, A.K. and Hollaway, L. (1985) The analysis of elastic-plastic adhesive stress in bonded lap joints in FRP structures. *Composite Structures*, **4** 135-160.

Press, W.H., Teukolsky, S.A., Vetterling, W.T. and Flannery, B.P. (1992) *Numerical Recipes in Fortran: The Art of Scientific Computing*. 2nd Edition, Cambridge University Press, UK, 751-778.

Raju, I.S. (1987) Calculation of strain-energy release rates with higher order and singular finite elements. *Eng. Frac. Mech.*, **28** 251-274.

Raju, I.S., Crews, J.H.J. and Aminpour, M.A. (1988) Convergence of strain energy release rate components for edge-delaminated composite laminates. *Eng. Frac. Mech.*, **30** 383-396.

Reeder, J.R. (1993) A bilinear failure criterion for mixed-mode delamination. *Composite Materials: Testing and Design*, *ASTM STP,* **1206 11** 303-322.

Reeder, J. R. & Crews, Jr, J. H. (1990) The mixed -mode bending method for delamination testing. *AIAA J.*, **28** 1270-1276.

Reeder, J.R. and Crews, J.H. (1991) Nonlinear analysis and redesign of the mixed-mode bending delamination test. NASA TM 102777, Hampton, VA, National Aeronautics and Astronautics Administration.

Renton, W. J. and Vinson, J. R. (1977) Analysis of adhesively bonded joints between panels of composite materials. *ASME J. of Applied Mechanics*, March 101-106.

Rice, J.R. and Sih, G.C. (1965) Plane problems of cracks in dissimilar media. *ASME J. of Applied Mechanics*, **32** 403-410.

Rice, J. R. (1968) A path independent integral and the approximate analysis of strain concentration by notches and crack. *ASME J. of Applied Mechanics,*. **35** 379-386.

Richardson, G., Crocombe, A.D. and Smith, P.A. (1993) A comparison of two- and three-dimensional finite element analysis of adhesive joints. *Int. J. Adhesion and Adhesives*, **13** (3) 193-200.

Ripling, E.J., Mostovoy, S. and Corten, H. (1971) Fracture mechanics: a tool for evaluating structural adhesives. *J. of Adhesion*, **3** 107-123.

Ripling, E.J., Santner, J.S. and Crosley, P.B. (1983) Fracture toughness of composite adherends, adhesive joints under mixed-mode I and III loading. *J. of Material Science*, **18** 2274-2282.

Ripling, E.J., Crosley, P.B. and Johnson, W.S. (1988) A comparison of pure mode I and mixed mode I-III cracking of an adhesive containing an open knit cloth carrier. *Adhesively Bonded Joints: Testing, Analysis and Design, ASTM STP,* **981**, W.S. Johnson, Ed., American Society for Testing and Materials, Philadelphia, pp.163-182.

Rispler, A.R. (1998) Analysis and optimization of highly loaded joints, PhD Thesis, department of Aeronautical Engineering, University of Sydney, Australia

Rispler, A.R., Steven, G.P. and Tong, L. (1997) Failure analysis of composite T-joints including inserts. *J. of Reinforced Plastics and Composites*, **16** (18) 1642-1658.

Roberts, T.M. (1989) Shear and normal stresses in adhesive joints. *J. Engineering Mechanics*, **115** (11) 2460-2479.

Rossettos, J.N. and Zhang E. (1993) On the peak shear stresses in adhesive joints with voids. *ASME J of Applied Mechanics*, **60** (June) 559-560.

Rowlands, R.E. (1985) Strength (failure) theories and their experimental correlation. Chapter 2, *Handbook of Composites - Failure Mechanics of Composites*, Edited by G. C. Sih and A. M. Skudra, Elsevier Science Publishers, **3** 71-125.

Rozvany, GN, Bendsoe, M.P. and Kirsh, U (1995) Layout optimization of structures. *Appl. Mech. Rev.* **48** (2) 41-119.

Rozvany, GIN (1989) Structural design via optimality criteria. Kluwer, Dordrecht, Holland

Russel, A. J. and Street, K. N. (1982) Moisture and temperature effects on the mixed-mode delamination fracture of unidirectional graphite/epoxy. *ASTM STP*, **876** 349-70.

Rybicki, E.F. and Kanninen, M.F. (1977) A finite element calculation of stress intensity factors by a modified crack closure integral. *Engineering Fracture Mechanics*, **9** (4) 931-938.

Saghizaden, H. and Dharan, C.K.H. (1986) Delamination fracture toughness of graphite and aramid epoxy composites. *ASME J. of Engineering Materials and Technology*, **108** (October) 290-295.

Sawyer, J.W. and Cooper P.A. (1981) Analytical and experimental results for bonded single lap joints with preformed adherends. *AIAA J.*, **19** (11) 1443-1451.

Sawyer, J.W. (1985) Effect of stitching on the strength of bonded composite single lap joints. *AIAA J.*, **23** 1744-1748.

Shenoi, R.A. and Hawkins, G.L. (1992) Influence of material and geometry variations on the behaviour of bonded tee connections in FRP ships. *Composites*, **23** 335-345.

Shenoi, R.A.and Violette, F.L.M. (1990) A study of structural composite tee joints in small boats. *J. of Composite Materials*, **24** 644-666.

Sheppard, A., Kelly, D.W. and Tong, L. (1998) A damage zone model for the failure analysis of adhesively bonded joints. *Int. J. of Adhesion and Adhesives*, **18** (6) 385-400

Shi, Y.P. and Cheng, S. (1993) Analysis of adhesive-bonded cylindrical lap joints subjected to axial load. *ASCE J. of Engineering Mechanics*, **119** (3) 584-602.

Sierakowski, R.L. and Newaz, G.M. (1995) *Damage tolerance in advanced composites*, Technomic Publishing Company, Lancaster, PA, USA.

Simo, J.C. and Ju, J.W. (1987) Strain- and stress-based continuum damage models - I. Formulation. *Int. J. of Solids and Structures*, **23** (7) 821-840.

Sneddon, I.N. (1961) The distribution of stress in adhesive joints. in *Adhesion* (Eley, D.D Ed.), Chapter 9, Oxford University Press.

Spies, G.J. (1953) The peeling of adhesive bonded metal. *J. Aircraft Engng.*, **25** 64.

Springer, G.S. and Finn, S.R. (1991) *Composite Plates Impact Damage, An Atlas*, Technomic Publishing Co. Inc. USA.

Sun, C.T. and Jih, C. J. (1987) On strain energy release rate for interfacial cracks in bimaterial media. *Eng. Frac. Mech.*, **28** 13-20.

Tada, Y and Ishikawa, T. (1989) Experimental evaluation of the effects of stitching on CFRP laminates with various shapes and loadings. *Key Engineering Materials*, **37** 305-316.

Theotokoglou, E.E. and Moan, T. (1996) Experimental and numerical study of composite T-joints. *J. of Composite Materials*, **30** (2) 190-209.

Thrall, E. W. (1979). Primary adhesively bonded structure technology (PABST): Design handbook for adhesive bonding. USAF Technical Report, AFFDL-TR-79-3119

Tong, L. (1994) Bond shear strength for adhesively bonded double lap joints. *Int. J. of Solids and Structures*, **31** (21) 2919-2931.

Tong, L., (1996) Bond strength of adhesive-bonded single lap joints. *Acta Mechanica*, **117** (1-4) 101-113.

Tong, L. (1997) An assessment of failure criteria to predict the strength of adhesively bonded composite double lap joints. *J. of Reinforced Plastics and Composites*, **16** (8) 698-713.

Tong, L. (1998a) Strength of adhesively bonded single-lap and shear-lap joints. *Int. J. of Solids and Structures*, **35** (20) 2601-2616.

Tong, L., (1998b) Failure of adhesively bonded composite single lap joints with embedded cracks. *AIAA Journal*, **36** (3) 448-456.

Tong, L. (1998c) Strength prediction of RTM single lap joints with transverse stitching. *ASME J. of Engineering Materials and Technology*, **120** (4) 300-303.

Tong, L., Cleaves, K., Kruckenberg, T. et al (1998) Effect of transverse stitching on the fracture loads of RTM single lap joints. *Key Engineering Materials*, **137** 195-202.

Tong, L. and Jain, L.K. (1995) Analysis of adhesive bonded composite lap joints with transverse stitching. *Applied Composite Materials*, **2** (6) 343-365.

Tong, L., Jain, L.K. and Leong, K.H. et al (1998) Failure of transversely stitched RTM lap joints. *Composite Science and Technology*, **58** (2) 221-227.

Tong, L., Kurrupu, M. and Kelly, D. (1997) Analysis of adhesively bonded composite double lap joints. *J. of Thermoplastic Composite Materials*, **10** (1) 61-75.

Tong, L., Sheppard, A. and Kelly, D. (1994) A numerical study of adhesively bonded composite panel-flange joints. *Composite Structures*, **28** (4) 449-458.

Tong, L., Sheppard, A. and Kelly, D. (1995) Relationship between surface displacement and adhesive peel stress in bonded double lap joints. *Int. J. of Adhesion and Adhesives*, **15** (1) 43-48.

Tong, L., Sheppard, A. and Kelly, D. (1996) The effect of adherend alignment on the behaviour of adhesive bonded double lap joints. *Int. J. of Adhesion and Adhesive*, **16** (4) 241-247.

Tong, L., Sheppard, A. Kelly, D. et al. (1998) Effect of joint flexibility in adhesively bonded composite panel-flange. *Composites Part B: Engineering*, **29** (3) 287-298.

Tong, L. and Steven, G.P. (1996) A numerical study on stresses in RTM lap joints reinforced with transverse stitching. *Composite Structures*, **36** (1-2) 91-100.

Tong, L., Zhu, X., and Steven, G.P. (1998) Damage tolerance of adhesively bonded composite joints. *Key Engineering Materials*, **145-149** 537-542.

Tong, L. and Zhuang, W.Z.L. (1997) Holographic measurement of the relative surface normal displacements in adhesive-bonded doubler. *J. of Adhesion Science and Technology*, **11** (7) 971-977.

Tsai M.-Y. and Morton, J. (1993) Mechanics of the adhesive-bonded single-lap joint specimen, Conference Proceedings: Pacific Int. Conference on Aerospace and Technology, Vol. II, PICAST'1 1993, December 1993, Taiwan, pp.936-944.

Tsai, M.Y. and Morton, J. (1994) An evaluation of analytical and numerical solutions to the single-lap joint, *Int. J. of Solids and Structures*, **31** (18) 2537-2563.

Tuttle, M.E., Barthelemy, B.M. and Brison, H.F. (1984) Strain measurement within a single-lap joint using embedded strain gages, *Experimental Mechanics*, **8** 31.

Vallat, M.F., Martz, P., Fontaine, I. and Schultz, S. (1986) The application of coherent optics to the study of adhesive joints. I. speckle photography. *J. of Applied Polymer Science*, **31** 309-321.

Vanderplaats, G.N. (1984) *Numerical Optimization Techniques for Engineering Design: with Application*, McGraw-Hill, New York.

Volkersen, O. (1938) Die Niektraftverteiling in Zugbeansprunchten mit Konstanten Laschenquerscritten. *Luftfahrforsch*, **15** 41-47.

Volkersen, O. (1965) Recherches sur la theorie des assemblages colles. *Construction Metallique*, **4** pp 3-13.

Wang, C.H. and Rose, L.F.R. (1997) Determination of triaxial stresses in bonded joints. *Int. J. Adhesion and Adhesives,* **17** (1) 17-25.

Williams, M.L. (1959) The stresses around a fault or crack in dissimilar media. *Bull. Seismol. Soc. America*, **49** 199-204.

Williams, J.G. (1988) On the calculation of energy release rates for cracked laminates. *Int. J. of Fracture*, **36** 101-119.

Wooley, G.R. and Carver, D.R. (1971) Stress concentration factors for bonded lap joints. *J. Aircraft,* **8** 817-820.

Xie, M. and Steven, G.P. (1997) *Evolutionary Structural Optimisation*, Springer-Verlag.

Yuceoglu, U. and Updike, D.P. (1980) Stress analysis of bonded plates and joints. *ASCE J. of Engineering Mechanics*, **106** (EM1) 37-56.

Zhou, H. M. and Rao, M.D. (1993) Viscoelastic analysis of bonded tubular joints under torsion. *Int. J. of Solids and Structures*, **30** (16) 2199-2211.

Zhuang, W.Z.L., Baird, J.P., Clark, R.K., Williamson, H.W. and Heslehurst, R.B. (1994) Peel displacement measurement of bonded double lap joints by holographic interferometry. Research report, Department of Aerospace and Mechanical Engineering, Australian Defence Force Academy, Canberra, 1994

Zienkiewicz, O.C. and Taylor, R.L. (1989) The Finite Element Method. 4th Edition. McGraw-Hill Book Company, England.

Index